IV International Scientific and Technical Conference "Metrology of Physicochemical Measurements" 2019

Journal of Physics: Conference Series Volume 1420

Suzdal, Russia
17 – 19 September 2019

ISBN: 978-1-7138-1325-5
ISSN: 1742-6588

Printed from e-media with permission by:

Curran Associates, Inc.
57 Morehouse Lane
Red Hook, NY 12571

proceedings.com

Some format issues inherent in the e-media version may also appear in this print version.

Institute of Physics
Dirac House, Temple Back
Bristol BS1 6BE UK

Phone: 44 1 17 929 7481
Fax: 44 1 17 920 0979

techtracking@iop.org

Additional copies of this publication are available from:

Curran Associates, Inc.
57 Morehouse Lane
Red Hook, NY 12571 USA
Phone: 845-758-0400
Fax: 845-758-2633
Email: curran@proceedings.com
Web: www.proceedings.com

TABLE OF CONTENTS

The IV International Scientific and Technical Conference "Metrology on Physicochemical Measurements" was held 17-19 September 2019 in Suzdal, Russia. It was organized by the All-Russian Scientific Research Institute of Physicotechnical and Radio Engineering Measurements (VNIIFTRI) with participation of the Cooperation on International Traceability in Analytical Chemistry (CITAC) under the slogan "Let's speak about measurement uncertainty".

About 100 participants from metrological, academic and high-education institutes, metrology service centers and other organizations of Belarus, Bulgaria, China, Israel, Germany, Switzerland and Russia took part in the Conference. There were 44 oral presentations, some posters and a small exhibition.

The Conference program included scientific and technical topics from the fields of electrochemistry, spectrometry, chromatography and other instrumental methods of physicochemical analysis; reference materials; measurements of dispersed parameters of particles in heterogeneous media, including aerosols and suspensions and air ions; applied and legal metrology and others.

During the framework of the Conference, interesting suggestions were made to further improve the level and quality of the metrological work performed, current problems in the field of ensuring the uniformity of measurements were highlighted, the opportunity to exchange experiences, establish contacts between organizations was provided. New achievements in the field of fundamental and applied research in the field of metrological support of physico-chemical measurements, world and domestic achievements in the development of measuring equipment in this field are presented.

After welcoming remarks by Prof.Dr.Sergey Donchenko, General Director of VNIIFTRI (Russia), such well-known scientists as Prof.Dr.Alfred Wiedensohler, Head of Experimental Aerosol and Cloud Microphysics Department (Leibniz Institute for Tropospheric Research, Germany); Dr. Samuel Wunderli, CEO “atomolmetron” GmbH, Independent Consultant on Metrology, CITAC member (Switzerland); Dr. Ilya Kuselman, Independent Consultant on Metrology, Editor of CITAC News and Chair of the IUPAC Interdivisional Working Party for Harmonization of Metrology and Quality Guidelines (Israel); Dr. Anatoly Trubitsyn, Scientific Advisor of Gorniy TsOT Ltd. (Russia); Dr. Natalia Muravskaya, Head of QM, VNIIOFI, Chair of COOMET TC 3.1”Quality Forum” (Russia); Dr. Anastasia Semenova, Deputy director of the Institute of Food Systems after Gorbatov (Russia) and e.g., as well as many young scientists took part in the conference with reports. In total, more than 40 reports were presented at the conference, of which 16 reports were made by scientists of VNIIFTRI. The conference was very active, heated discussions continued during the breaks.

The participants were satisfied by the scientific level and usefulness of the Conference.

The next such conference is planned by VNIIFTRI for 14-16 September 2021 in beautiful historical place - Park-Hotel "MOROZOVKA", Moscow region, Mendeleevo, Russia.

Journal of Physics: Conference Series **1420** (2019) 011001 doi:10.1088/1742-6596/1420/1/011001

The Scientific secretary of the Conference *Dr. Narine G Oganyan*

Journal of Physics: Conference Series **1420** (2019) 011002 doi:10.1088/1742-6596/1420/1/011002

IV INTERNATIONAL SCIENTIFIC AND TECHNICAL CONFERENCE «METROLOGY OF PHYSICAL AND CHEMICAL MEASUREMENTS»

17-19 September 2019, Suzdal, Russia

PROGRAM COMMITTEE	
CHAIR	
Dr. Sergey Golubev	Deputy Head of ROSSTANDART, Vice-President of COOMET
CO-CHAIR	
Dr., Prof. Sergey Donchenko	General Director of the State Metrology Institute "VNIIFTRI"
DEPUTY CHAIR	
Dr. Vladimir Dobrovolskiy	Head of Physicochemical and Electrical measurements Department, VNIIFTRI, Secretary of OIML TC17/SC3 "pH-metry"
MEMBERS	
Dr. Wofram Bremzer	Department of Analytical Chemistry, Federal Institute for Materials Research and Testing (BAM for VDI/VDE), Germany
Dr. Ilya Kuselman	Independent Consultant on Metrology, Editor of CITAC News and Chair of the IUPAC Interdivisional Working Party for Harmonization of Metrology and Quality Guidelines, Israel
Prof. Dr. Alfred Wiedensohler	Department Head, Experimental Aerosol and Cloud Microphysics, Leipniz Institute for Tropospheric Research, Germany
Dr. Samuel Wunderli	CEO Consulting Committee "atomolmetron" GmbH, Independent Consultant on Metrology, Switzerland, CITAC member
Prof. Li Zhaojun	Department Head, Institute of Process Engineering, China
Dr. Andrey Baturin	Director of the State Metrology Institute "VNIIOFI"
Dr. Dmitriy Gladilovich	Deputy Director, LUMEX Marketing LLC, Analytical Instrumentation
Dr. Aleksandr Komarov	Deputy Director, Institute VGNKI
Dr. Elene Kulyabina	Head of Laboratory, VNIIMC
Dr. Aleksandr Kuzin	Director of the State Metrology Institute «VNIIMC»
Dr. Sergey Medvedevskiy	Director of the State Ural Institute for Metrology «UNIIM»
Dr. Natalya Muravskaya	Head of QM, VNIIOFI, Chair of COOMET TC 3.1"Quality Forum"

Journal of Physics: Conference Series **1420** (2019) 011002 doi:10.1088/1742-6596/1420/1/011002

Dr. Narine Oganyan	Deputy Head of Physicochemical and Electrical measurements Department, VNIIFTRI, CITAC member
Dr. Anton Pronin	Acting Director of the D.I. Mendeleev State Institute for Metrology VNIIM
Dr. Egor Sobina	Deputy Director of UNIIM
Dr. Anastasia Semenova	State Scientific Center on food systems after V.M. Gorbatov

Peer review statement

All papers published in this volume of *Journal of Physics: Conference Series* have been peer reviewed through processes administered by the proceedings Editors. Reviews were conducted by expert referees to the professional and scientific standards expected of a proceedings journal published by IOP Publishing.

Journal of Physics: Conference Series **1420** (2019) 012001 doi:10.1088/1742-6596/1420/1/012001

How many shades of grey are in conformity assessment due to measurement uncertainty?

Ilya Kuselman[1], Francesca R Pennecchi[2], Ricardo J N B da Silva[3], D Brynn Hibbert[4]

[1] Independent Consultant on Metrology, 4/6 Yarehim St., 7176419 Modiin, Israel
[2] Istituto Nazionale di Ricerca Metrologica (INRIM), Strada delle Cacce 91, 10135 Turin, Italy
[3] Centro de Química Estrutural, Faculdade de Ciências da Universidade de Lisboa, Edifício C8, Campo Grande, 1749-016 Lisboa, Portugal
[4] School of Chemistry, UNSW Sydney, Sydney NSW 2052, Australia

Abstract: When a measured value of a property of a material or object differs from the upper or lower specification limit (actual or 'true' value) by the expanded measurement uncertainty or more, there is the clear decision on the material conformity or nonconformity - 'white' or 'black'. In the interval from the measured value to the specification limit, covered by the expanded measurement uncertainty ('grey zone'), risks of false decisions on conformity increase. Several kinds of the risks, named 'shades of grey', should be taken into account. For a multicomponent material there are four kinds of particular risks for each property value of the material (e.g. component concentration or content), and four kinds of total risks related to the material as a whole. Therefore, for $n > 1$ properties under control for the material conformity assessment one can distinguish $4(n+1)$ kinds of risks of false decisions – shades of grey.

Conformity assessment is the demonstration that specified requirements relating to a product, process, system, person or body are fulfilled [1]. Conformity of a product is assessed before it is placed on the market. The conformity assessment procedure for each product (e.g. a material) is specified in the applicable product legislation. Standard specifications for chemical composition of a multicomponent material – a medication, alloy, etc. – are tolerance limits of the actual ('true') concentration or content c_i of the i-th component, $i = 1, 2, \ldots, n$, including main components and impurities or groups of impurities. Conformity assessment of an item (material batch or lot) is based on comparing the measured concentration or content c_{im} with tolerance/specification limits. Since any c_{im} value has associated measurement uncertainty, acceptance limits for measurement results can be used in addition to tolerance limits. In these cases, the decision rules (does the test item conform or not?) are based on comparing the measured property values c_{im} with the acceptance limits [2]. The interval between a tolerance limit and corresponding acceptance limit is the 'grey zone', where probabilities of false decisions on conformity of the item are impermissible.

When tolerance limits have been defined by already taking into account measurement uncertainty, acceptance limits and tolerance limits coincide.

Several kinds of risk of a false decision on conformity of an item may be called *shades of grey*. The probability of accepting a batch of the material, when it should have been rejected, is the 'consumer's risk', whereas the probability of falsely rejecting the batch is the 'producer's risk'. For a specified batch, they are referred to as the 'specific consumer's risk' and the 'specific producer's risk' R_{ci}^*, respectively, for the i-th particular component of the material under control. The risks of incorrect conformity assessment of a batch randomly drawn from a statistical population of such batches are the 'global consumer's risk' and the 'global producer's risk' R_{ci}, as they characterize the material production globally [2]. If a tolerance limit and corresponding acceptance limit coincide, the grey zone collapses, however the risks are still above zero at any measurement result. Thus, there are four shades of grey for each property value of an item – concentration or content of i-th particular component of a material (consumer's and producer's risks, both are specific and global).

In general, a component-by-component evaluation of the risks of a material conformity assessment is not complete, as it does not give an answer to the question of the probability of a false decision on conformity of the material as a whole. When conformity assessment for each i-th component of a material is successful (i.e. the particular specific R_{ci}^* or global R_{ci} risks are small enough), the total probability of a false decision concerning the material as a whole (the *total* specific R_{total}^* or *total* global R_{total} risk) might still be significant. Evaluation of the total risks when test/measurement results are not correlated is detailed in our publications on conformity assessment of denatured alcohols [3] and total suspended particulate matter in ambient air [4]. Discussion of the cases of correlated data is available in the papers on conformity assessment of a medication [5] and an alloy [6]. These evaluations are based on the Bayesian approach [2] and performed in R programming environment. In paper [6] we have proposed also a solution of the inverse problem: the use of total specific risk values for setting multivariate acceptance limits. In papers [7, 8] tutorial and user-friendly MS Excel spreadsheets for Bayesian evaluation of total specific and global risks, respectively, are described.

Hence, there are four kinds of particular risks for each i-th property value (component concentration or content) of a material, and four kinds of total risks. Therefore, for $n > 1$ components under control one can distinguish $4(n+1)$ kinds of risks of false decisions – shades of grey. For example, for two components this means - 12, for three components – 16, and for four components – 20 shades of grey.

References

[1] ISO/IEC 17000:2004. Conformity assessment. Vocabulary and general principles
[2] JCGM 106:2012. Evaluation of Measurement Data – The Role of Measurement Uncertainty in Conformity Assessment
[3] Kuselman I, Pennecchi F R, da Silva R J N B and Hibbert D B 2017 Talanta **164** 189
[4] Pennecchi F R, Kuselman I, da Silva R J N B and Hibbert D B 2018 Chemosphere **202** 165
[5] Kuselman I, Pennecchi F R, da Silva R J N B and Hibbert D B 2017 Talanta **174** 789
[6] Kuselman I, Pennecchi F R, da Silva R J N B, Hibbert D B and Anchutina E 2018 Talanta **189** 666
[7] da Silva R J N B, Pennecchi F R, Hibbert D B and Kuselman I 2018 Chemometrics and Intelligent Laboratory Systems **182** 109
[8] da Silva R J N B, Lourenço F R, Pennecchi F R, Hibbert D B and Kuselman I 2019 Chemometrics and Intelligent Laboratory Systems **188** 1

Journal of Physics: Conference Series **1420** (2019) 012002 doi:10.1088/1742-6596/1420/1/012002

Uncertainty evaluation in general including simple artificial applications and Monte Carlo uncertainty evaluation

Samuel Wunderli

dipl.chem.ETH / Dr.sc.nat.ETH, Independent Consultant on Metrology, Director & Owner of atomolmetrion GmbH, Chemin de la Binna 10, 1786 Sugiez, Switzerland

e-mail : xiongmao@gmx.ch phone +41 (0)26 673 06 19

Abstract : A short introduction of the general bottom-up uncertainty evaluation approach according to the ISO-GUM, JCGM 100 was presented [1, 3]. The Monte-Carlo Method (MCM) for the evaluation of the measurement uncertainty was outlined [2]. An application of the Monte-Carlo technique to the standard addition method using a simple excel-spreadsheet was explained in order to encourage the use of this important and powerful approach. A four and a five point standard addition method were compared for different sets of uncertainties of input variables and represented graphically.

Measurement uncertainty evaluation using the bottom-up approach method outlined in the ISO-GUM (identical to JCGM 100) and recommended by a number of leading international member organisations is now in use successfully for about a generation in all fields of measurements. The recently revised laboratory accreditation standard ISO/IEC 17025:2017 again requires that laboratories evaluate the measurement uncertainty for their reported measurement values [4]. In order to extend the use of the ISO-GUM the Joint Committee for Guides in Metrology (JCGM) under the guidance of the BIPM (Bureau International des Poids et Mesures, Paris) and its member organisations have published the supplement 1 [2], which explains the use of the Monte Carlo Method to evaluate the measurement uncertainty. It is especially useful as the method gives access to the distribution of the measurand values. No partial derivatives have to be calculated. In case of a non-linear model equation for the measurand the method is very much to be recommended as the full information of the input variables is reflected in the distribution of the measurand values and not only a the most probable value for the measurand and the corresponding combined standard uncertainty.

This contribution focusses on giving some detailed practical information on how to perform MCM using excel. The random number generator and also the number of possible sampling points in excel have been improved by Microsoft and are now fit for purpose, which was not the case in the past.

A procedure to perform MCM using excel for uncertainty evaluation is outlined as follows [5] :

1. The model equation for the measurand needs to be created using all available knowledge about the measurement system. This is generally the most demanding task.
2. For each input quantity (either type A or type B) in the model equation we need either a measured or an assumed probability density function characterised by the defining parameters for the distributions. Most commonly these distributions are the rectangular, symmetric triangular and the Gaussian distributions. The above mentioned distributions are given by two

parameters wich are the central value and the second moment (standard deviation). For further distributions see chapter 6.4 in JCGM 101.

3. If we have to consider a bias (sum of all systematic influences) our model equation needs a proper adaption. The bias is a value with a sign and has a corresponding uncertainty. Mostly the bias correction can be characterised as a simple distribution with a value and a standard deviation. The measurand is corrected for if the bias is large enough.
4. For each of the input quantities x_j we calculate a simulated random value $x_{i,j}$ that reflects the distribution characteristics of that input quantity x_j. From the set of these realisations $x_{i,j}$ of each input quantity x_j we now calculate the values y_i of the measurand and repeat this procedure for say 100'000 times. JCGM 101 gives advice on the number of simulations (chapter 7.2 of JCGM 101). Now we have a vector y_i for the measurand with all realisations from the many simulations i.

Sampling from a **rectangular distribution** with limits a_r,b_r :

$$x_{i,j} = m_{r,j} - \frac{r_{r,j}}{2} + RAND() \cdot r_{r,j} = a_{r,j} + RAND() \cdot \frac{\left[b_{r,j} - a_{r,j}\right]}{2}$$

i: MCM simulation number, j: input quantity number

$m_{r,j}$: mean value of rectangular distribution, $RAND()$: random number $[0,1]$

$r_{r,j} = \left[b_{r,j} - a_{r,j}\right]$: range

The excel function *RAND*() produces a 15 digit random number in the range [0,1]. The values a_r, b_r characterise the minimum and the maximum of the rectangular distribution with the mean value $0.5 \cdot (a_r + b_r)$.

Sampling from a **symmetric triangular** distribution with limits a_t,b_t:

$$x_{i,j} = m_{t,j} - \frac{r_{t,j}}{2} + \left\{RAND() + RAND()\right\} \cdot \frac{r_{t,j}}{2} = a_{t,j} + \left\{RAND() + RAND()\right\} \cdot \frac{\left[b_{t,j} - a_{t,j}\right]}{2}$$

Whenever the random number function is called a new value is created. That means the above two randon number functions are independently evaluated with generally different numbers.

Sampling from a **Gaussian distribution** sampling:

$$x_{i,j} = NORMINV\left\{RAND(), m_{g,j}, s_{g,j}\right\}$$

$m_{g,j}$ mean of Gaussian distribution of input variable j

$s_{g,j}$ standard deviation of Gaussian distribution of input variable j

5. These simulated values y_i of the measurand Y have to be analysed to get the values for the measurand expectation and the expanded measurement uncertainty. The mean value is the expectation value for the measurand. Two times to calculated standard deviation is nearly identical to the expanded measurement uncertainty. Mostly the distribution is not totally symmetric and also not perfectly Gaussian. Therefore we can use the calculated, cumulative probability distribution where we have to extract the 2.5% value and the 97.5% value to get the 95% range, that we attribute to the expanded measurement uncertainty.
6. The cumulative probability distribution is calculated as follows :

a) First create a class vector with the numbers 1 to the maximum class number (e.g. 1 to 100). Calculate the values of the classes using the following expression repeatedly over all class numbers:

$$value(class)_i = \min\left(y_{simul}\right) + \left[N^o class_i - 1\right] \cdot \frac{1}{\left(N^o class_{\max} - 1\right)} \cdot \left\{\max\left(y_{simul}\right) - \min\left(y_{simul}\right)\right\}$$

The class values will start from the minimum of the simulated values ($\min(y_{\text{simul}})$) and the values ($value(class)_i$) with (total number $N°class_{max}$) will be evenly distributed between the minimum and the maxiumum of the simulated y_i values.

b) From the simulated values for the measurand y_i we count the frequency (that is the number of values in each class) for each of the said hundred equally distributed classes using the following excel-function :
FREQUENCY{*y*1 : *y*100000 ; class1 : class100}
(for e.g. 100'000 simulations and 100 classes in which *y*1 is the field of the first simulated value and *y*100000 is the field of the last simulated values and class1 the field containing the value for the first class and class100 the field containing the value for the last class.).
The *FREQUENCY* function has two arguments : the vector of the simulated measurand values (*y*1 : *y*100000) and the vector for the calculated class values. Here we chose as an example 100 classes. The function returns a vector and it must be entered using CTRL+shift+enter. Corrections are made by pressing F2, edit>erase-all. The use of the *FREQUENCY*-function needs some training.

c) From the number of values in each class the cumulative frequency is calculated for each class by adding the number of values for all preceeding classes. Subsequently also the precentage values of the cumulative frequency are calculated.

d) From the cumulative frequency distribution the expanded measurement uncertainty U is evaluated at the 2.5% and the 97.5% cumulative values.

7. The data are displayed graphically including the evaluated numbers for the expectation value and the expanded measurement uncertainty. Any change in the input quantity parameters are directly displayed in the excel table and the corresponding graph.

The MCM has been applied to the evaluation of the expanded measurement uncertainty estimation for an analyte value measured using the standard addition method, which is very common in analytical chemistry. By changing the input parameters the required expanded uncertainty can be simulated.
For the five point calibration four samples have known amounts of analyte added.
For high precision this is best done using a gravimetric approach and not a volumentric one.

The calibration curve is a linear function (eq. 1). Apart form the sample without added standard four additional samples with added known amounts of an analyte standard are measured (Fig. 1).

$$y = f(x) = a + b \cdot x$$

linear function

$$a = \frac{\sum_{i=1}^{n} y_i \cdot \sum_{i=1}^{n} x_i^2 - \left(\sum_{i=1}^{n} x_i \cdot y_i\right) \cdot \sum_{i=1}^{n} x_i}{n \cdot \sum_{i=1}^{n} x_i^2 - \left(\sum_{i=1}^{n} x_i\right)^2} \qquad b = \frac{\overline{y} - a}{\overline{x}} \qquad \text{(eq 1)}$$

Best linear function from experimental values x_i, y_i.

Uncertainties in x_i and y_i are taken into account. Gaussian distributions are used for all input variables using the sampling procedure described under point 4, above.

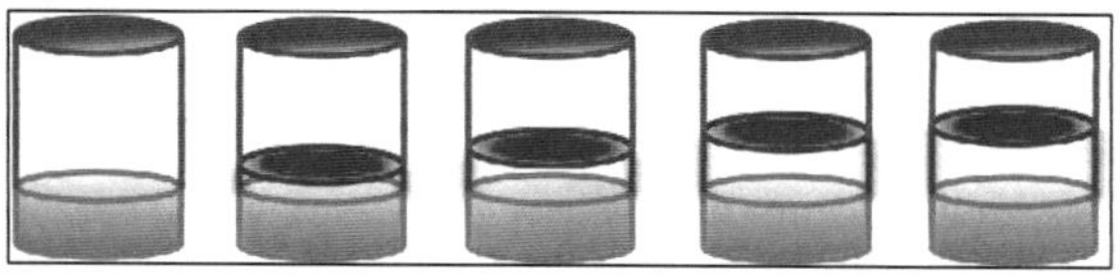

Figure 1. One sample without and four samples with known amounts of the analyte standard are used for the five point standard addition method.

Comparison of calibration, 4 and 5 points (3 and 4 additions)
Standard uncertainties for x_i are 1 % or less, for y_i about 5% or less

4 points

x, mg/L	*y*	*u*(*x*), *mg/L*	*u*(*y*)
0	0.38	5.77E-03	0.02
1.98	0.48	0.02	0.02
3.96	0.54	0.02	0.02
5.94	0.59	0.02	0.02

y	11.6	mg/L
U	3.07	mg/L
low	8.5	mg/L
high	16.4	mg/L
(*k* = 2, 95%)		

rel. expanded uncertainty 27 %

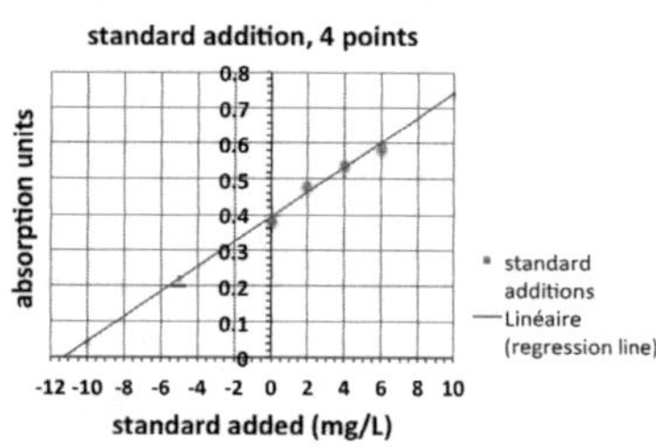

5 points

x , mg/L	*y*	*u*(*x*), mg/L	*u*(*y*)
0	0.38	5.77E-03	0.02
1.98	0.48	0.02	0.02
3.96	0.54	0.02	0.02
5.94	0.59	0.02	0.02
7.12	0.65	0.02	0.02
			0.02

y	11.2	mg/L
U	2.3	mg/L
low	8.7	mg/L
high	14.6	mg/L
(*k* = 2, 95%)		

rel. expanded uncertainty: 21 %

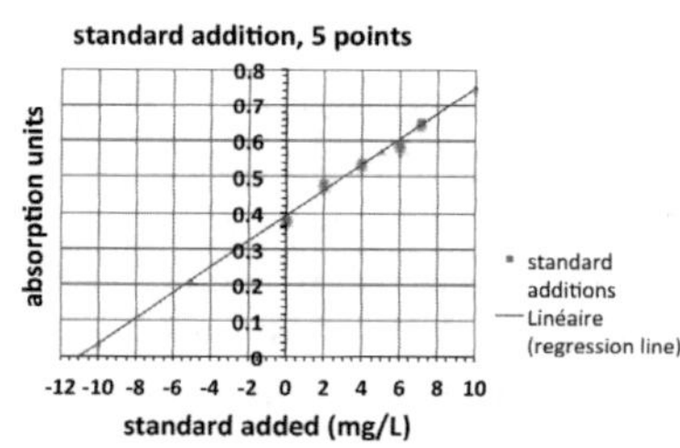

Tables 1,2 & Figures 2,3. Values and uncertainties for a 4-point and a 5-point standard addition.

Measurand values and expanded uncertainties estimated by MCM using Gaussian distributions for all input quantities. The expectation value for the measurand is the intersection point between the calibration line and the concentration axis.

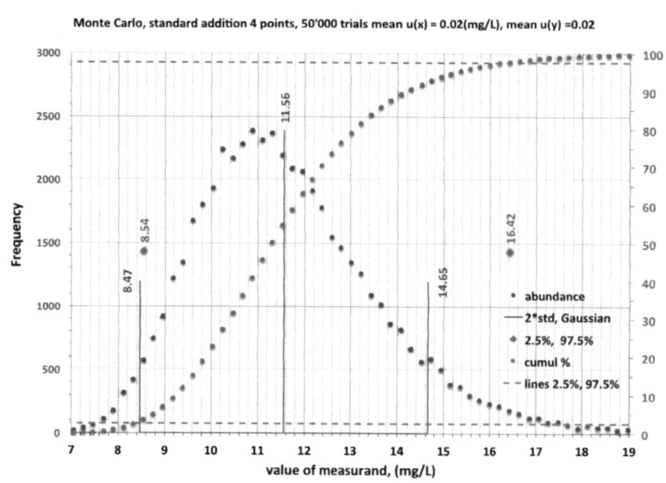

Figure 4. Graphical display of the measurand distribution with a 4-point standard addition and the corresponding cumulative frequency distribution.
Uncertainties improved for the x_i and y_i values compared to Fig.2,3.

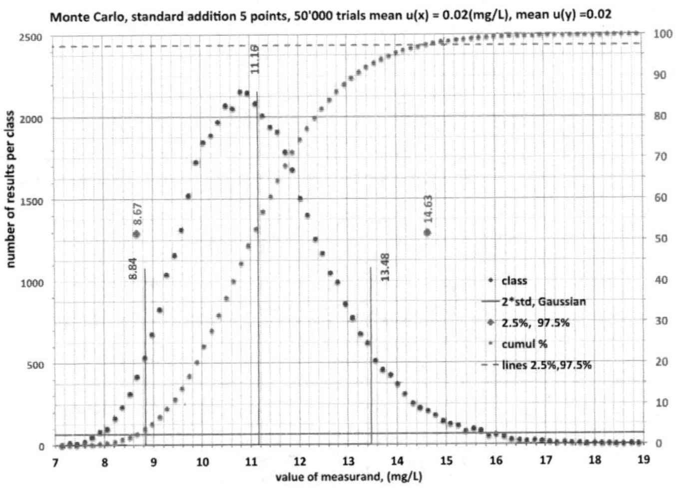

Figure 5. Graphical display of the measurand distribution with a 5-point standard addition and the corresponding cumulative frequency distribution. Uncertainties improved for the x_i and y_i values compared to Fig.1,2.

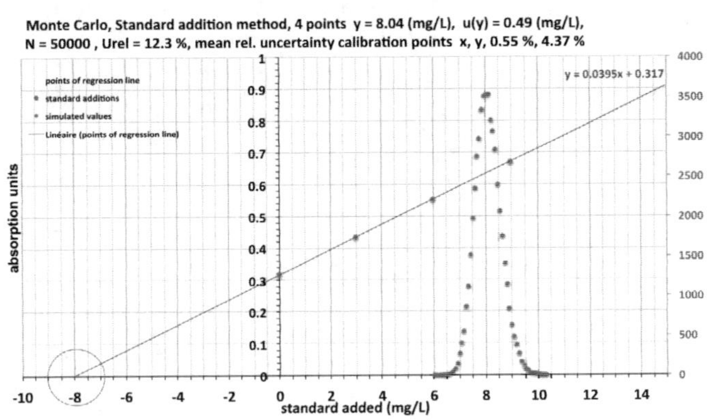

Figure 6. Graphical display of the measurand distribution with a 4-point standard addition and the corresponding cumulative frequency distribution.
Values for input quantities are different from Fig.1 to 5. Compare result only with Fig. 7.

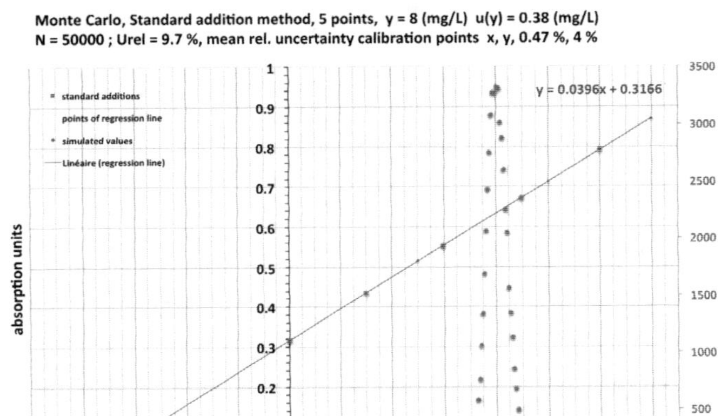

Figure 7. Graphical display of the measurand distribution with a 5-point standard addition and the corresponding cumulative frequency distribution.
Values for input quantities are different from Fig.1 to 5. Compare result only with Fig. 6. As expected a five point calibration reduces the expanded measurement uncertainty (cf. distribution of measurand in green).

Further improvement can be seen if the uncertainties of both x_i and y_i values are reduced

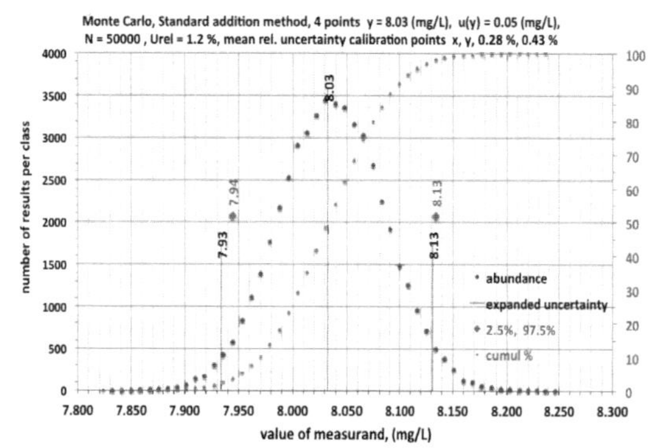

Figure 8. Further reduction of the uncertainty of the input quantity values reduces the expanded measurement uncertainty.
To achieve an expanded uncertainty of 1.2% for the measurand the mean relative uncertainties of the input quantity values have to be reduced to 0.25% and 0.43% (cf. title of the Fig.8), which is rather demanding. Compare Fig. 8 with Fig.9. In blue : calculated mean and expanded standard deviation (2 x std). In green : Expanded uncertainty from evaluation of cumulative frequency distribution.

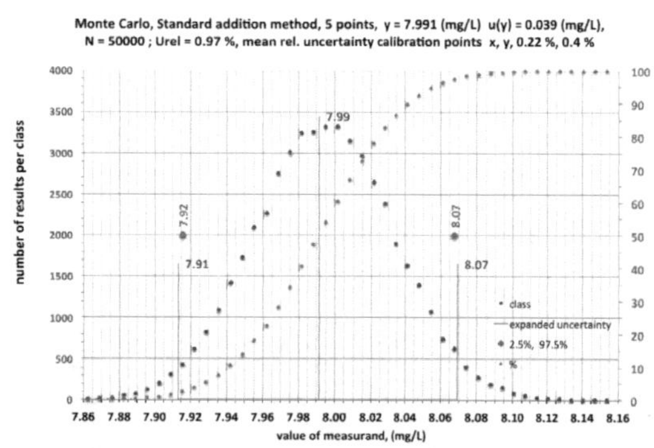

Figure 9. Further reduction of the uncertainty of the input quantity values reduces the expanded measurement uncertainty.
To achieve an expanded uncertainty of 0.97% for the measurand the mean relative uncertainties of the input quantity values have to be reduced to 0.22% and 0.4% (cf. title of the Fig.9), which is rather demanding. Compare Fig. 9 with Fig.8. In blue : calculated mean and expanded standard deviation (2 x std). In green : Expanded uncertainty from evaluation of cumulative frequency distribution.

Expressions used often for MCM (see JCGM 101, p.19 ff)

rectangular distribution: JCGM 101, Subclause 6.4.2.4

$$x_r = m - \frac{r}{2} + z \cdot r = m - \frac{r}{2} + RAND() \cdot r$$

m mean, r range, z random variable: $RAND()$ $[0,1]$

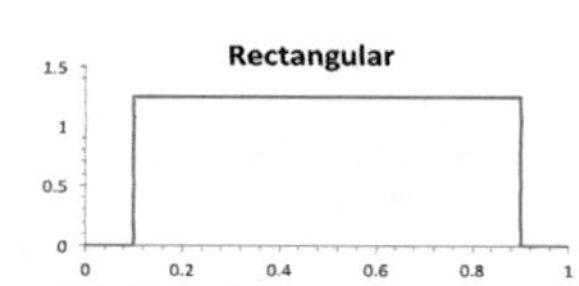

Gaussian distribution: JCGM 101, Subclause 6.4.7

$$x_g = NORMINV(z,\, m,\, std) = NORMINV\{RAND(), m, std\}$$

z random variable $[0,1]$, m mean, std standard deviation

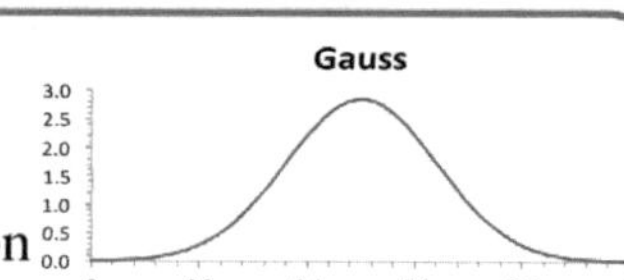

triangular symmetric distribution: JCGM 101, Subclause 6.4.5.4

$$x_t = m - \frac{r}{2} + (z_1 + z_2) \cdot \frac{r}{2} = m - \frac{r}{2} + \{RAND() + RAND()\} \cdot \frac{r}{2}$$

z_1, z_2 random variables $[0,1]$, m mean, r range

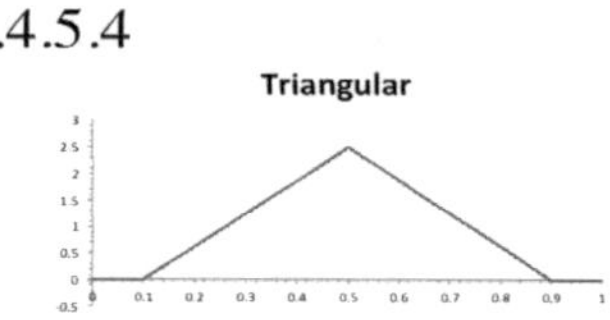

Figure 10. Most commonly used distribution for MCM sampling to evaluate the expanded measurement uncertainty. For further distributions see JCGM 101 [2], chapter 6.4.

References:

(all documents are accessible free of charge with the exception of [4]) :

[1] ISO-GUM, JCGM 100 :2008 (GUM 1995 with minor corrections),
[2] BIPM.JCGM 101 : 2008, Evaluation of measurement data – Supplement 1 to 'Guide to the expression of uncertainty in measurement' – Propagation of distributions using a Monte Carlo method., BIPM.
[3] EURACHEM/CITAC Guide (2012), https://www.eurachem.org/index.php/publications/guides
[4] General requirements for the competence of testing and calibration laboratories, ISO/IEC 17025 :2017, Third edition 2017-11, ISO Geneva, Switzerland.
[5] Excel files are available from the author on request.

Journal of Physics: Conference Series **1420** (2019) 012003 doi:10.1088/1742-6596/1420/1/012003

Measurement uncertainty and corresponding risk of false decisions

N G Oganyan

All-Russian Scientific Research Institute of Physicotechnical and Radio Engineering Measurements (VNIIFTRI), Mendeleevo, Moscow region, Russia, 141570

e-mail: oganyan@vniiftri.ru

Abstract. The work describes a relationship between risks of false decisions in conformity assessment and measurement uncertainty, shows that minimization of measurement uncertainty decreases the risks.

Measurement uncertainty and corresponding risk of false decisions are an integral part of human activity including economics, medicine, business, personal and social life, etc. Uncertainty exists objectively, regardless of our understanding.

A decision is making when the object under consideration is assessed for its compliance with the specified requirements (conforms or not). "Conformity assessment is demonstration that specified requirements relating to a product, process, system, person or body are fulfilled" [1].

However, like all other resources, information is limited. Therefore, most decisions are made under incomplete knowledge about an object and a consequence of this condition is the risk of false decisions. It is impossible to ignore today measurement uncertainty and the following risk. The reason for such a close relationship is the dependence of the decision making on completeness and reliability of the information received. Thus, in order to decrease the risks of making false decisions, it is necessary to expand the information about the object under testing, i.e. to minimize the measurement uncertainty.

Since practically all types of human activity are associated with measurements, including industrial, commercial, scientific, environmental, health and safety activities, measurement is considered as the main source of information on characteristics / parameters of the controlled object or subject (medicine).

When conformity assessment is based on measurement, a systematic examination of the object's compliance with the established requirements is performed. The aim of the conformity assessment is to ensure that the object complies with the established requirements [2].

According to the International Vocabulary of Metrology (VIM) [3], the aim of measurement is obtaining relevant information about a quantity of interest, the measurand (in analytical chemistry it is concentration or content of analyte). The more information we obtain on the measured quantity, the higher is the measurement quality.

The best estimate of the measurand is the measured quantity value and associated measurement uncertainty, which is defined as "non-negative parameter characterizing the dispersion of the quantity values being attributed to a measurand, based on the information used non-negative parameter characterized the variance of the value" [3]. Understanding the significance of existing problems in measurements, in 1993, seven international organizations jointly developed the "Evaluation of measurement data — Guide to the expression of uncertainty in measurement" (GUM) and the "International vocabulary of metrology - Basic and general concepts and associated terms" (VIM). Later, in 1997 a Joint Committee for Guides in Metrology (JCGM), chaired by the Director of the Bureau International des Poids et Mesures (BIPM), was created by that seven international organizations. In turn, the Joint Committee formed two Working Groups (WGs). WG 1, "Expression of uncertainty in measurement", has the task to promote the use of the GUM and to prepare Supplements and other documents for its broad application, in particular JCGM 106 guidelines for use of uncertainty in conformity assessment [4]. W G 2, "Working Group on International vocabulary of basic and general terms in metrology (VIM)", has the task to revise and promote the use of the VIM.

According to the developed documents, as a rule, the arithmetic mean of the observed quantity values is taken as the measured value and their standard deviation is used for evaluation of measurement uncertainty [3].

The national primary standard should have the highest measurement accuracy (the minimum uncertainty) in the country, according to the national infrastructure in the metrology field [5]. At the same time, the accuracy of measurements decreases during the transfer of the unit from the National Primary Standard to other standards and measuring instruments, in accordance with the hierarchy of measuring instruments [6], and, as a result, the uncertainty increases. This uncertainty increasing occurs regardless of whether a laboratory takes it into account or not. Knowledge the measurement procedure and the measurement uncertainty, it is possible to evaluate and minimize the risk of false decisions of compliance or non-compliance with the technical specifications or standards.

References.

[1] ISO/IEC 17000:2004 Conformity assessment - Vocabulary and general principles (IDT)
[2] ISO 10576-1:2003 Statistical methods -- Guidelines for the evaluation of conformity with specified requirements -- Part 1: General principles
[3] BIPM, IEC, IFCC, ILAC, ISO, IUPAC, IUPAP and OIML. International vocabulary of metrology - Basic and general concepts and associated terms (VIM), 3-dr edition, JCGM 100
[4] BIPM, IEC, IFCC, ILAC, ISO, IUPAC, IUPAP and OIML. Evaluation of measurement data - The role of measurement uncertainty in conformity assessment, JCGM 106
[5] OIML D1:2012. Considerations for a Law on Metrology
[6] OIML D5:1982 Principles for the establishment of hierarchy schemes for measuring instruments

Journal of Physics: Conference Series **1420** (2019) 012004 doi:10.1088/1742-6596/1420/1/012004

State and prospects for development of measurements of aerosol particle concentration, as well as air ions

V I Dobrovolskiy

All-Russian Scientific Research Institute of Physicotechnical and Radio Engineering Measurements (VNIIFTRI), Mendeleevo, Moscow region, Russia, 141570

e-mail: mera@vniiftri.ru

Abstract. The author highlighted the problems in Russia associated with the metrological support of measurements of the mass concentration of suspended particles of RM-10, RM-2.5, RM-1 fractions, as well as the number concentration of medium and heavy air ions.

Suspended particles are very dangerous for human health, as they are able to penetrate deep into the lungs and accumulate there.

As it is, in the Russian Federation there are hygienic normative document GN 2.1.6.3492-17 "Maximum permissible concentrations (MPC) of pollutants in the atmospheric air of urban settlements and rural area", which establish MPC of suspended particles size fractions PM-10 and PM-2.5 in the atmospheric air. Yet, there are no established norms of MPCs of ultrafine particles (PM-1), although they are the most harmful to human health due to their high penetrating ability and ability to be a suspension for a long time.

Besides uncharged solid particles, the PM-1 fraction also include medium and heavy air ions, the size of which is in the nanometer range. Currently, there is no metrological support for measurements of medium and heavy air ions concentration in Russia, although they have a strong effect on the human health.

One of the most advanced and actual tasks is metrological assurance of suspended particles mass concentration measurements in size fractions PM-10, PM-2.5, PM-1, as well as the medium and heavy air ions number concentration.

To solve the problems, laboratories of VNIIFTRI are actively study on these measurement areas.

Journal of Physics: Conference Series **1420** (2019) 012005 doi:10.1088/1742-6596/1420/1/012005

Spectral methods in physical and chemical measurements

N P Muravskaya

All-Russian Scientific Research Institute of Optical and Physical Measurements (VNIIOFI), Moscow, st. Ozernaya, 46, Russia

e-mail: vniiofi@vniiofi.ru

Abstract: The paper presents the spectral methods used in high-precision physico-chemical measurements.

Metrological support for measurements of liquid and solid chemical compositions is one of the most important problems of modern metrology. There are necessities of high-precision, reliable results of measurements the content of components in liquid and solid in such important industrial sectors, branches of science and technologies, as the chemical, nuclear, food, metallurgy, aviation, medicine, biology, ecology, agriculture and etc.

The problem of metrological support for measurements of chemical composition of substances became actual in the last century, since the most of reference materials of chemical composition of substances were absent. In practice, the analytical control was dominating. Components concentration of some chemical substances were determined by other chemical substances. In this case, the degree of the purity of substances and the error of the determined mass fraction of the main substance were commensurate. In the certificates of produced chemical reagents, the mass fraction of the main substance was indicated (and it is still indicated) as a limit “not less”. Such marking makes difficult and not possible in practice to use them for the quantitative chemical analysis.

High-precision measurement methods based on fundamental physics laws used exact mathematical equations and physical constants make it possible to establish the interrelation of measured units with the composition or properties of substances. These methods are based on the measurement of the SI units. This allows to almost completely eliminate the systematic errors associated with the procedure for the mutual determination of the component composition, when the determination of the component composition of some chemicals is carried out using others. The high-precision methods used for this purpose should have a good theoretical basis, be verified experimentally, considered to be primary, according to the modern terminology. Primary method of measurements are the methods with the highest metrological properties, in which models (mathematical equation) and implementation are fully described and understood and expressed in the units of SI.

Consultative Committee on the Quantity of Substance - Metrology in Chemistry (CCQM of BIPM) defines the following methods as primary:

- gravimetry
- coulometry
- titrimetry
- inductive-coupled plasma
- spectrometry of resonance line
- isotopic dilution mass spectrometry
- neutron activation analysis.

Since the use of spectral methods are at the one of the first places of study on substances and materials composition, the use of spectral measurements in the creation of new materials and technologies plays an important role in ensuring scientific and technological progress to achieve the high quality of products, to realize of innovation technologies, to save of materials and human sources, to increase the reliability and competitiveness, to be safety of modern technology and production, for environmental protection.

Development of primary reference measurement procedures for food quality control

Sergey V Medvedevskikh[1], Maria Y Medvedevskikh[1], Anna S Sergeeva[1], Vasilisa B. Baranovskaya[2]

[1]Ural Research Institute for Metrology, 4 Krasnoarmeyskaya, Ekaterinburg, Russia, 620000

e-mail: lab241@uniim.ru

[2]Kurnakov Institute of General and Inorganic Chemistry of the Russian Academy of Sciences 31 Leninsky Prospect, Moscow, Russia, 119991

Abstract. The work is about the development of primary reference measurement procedures for food quality control based on the methods of gravimetry, butymetry, photometry, chromatography, spectrometry and others.

The State verification schedule for measuring of water content of firm and loose materials was developed and approved in accordance with the Order of the Federal Agency for Technical Regulation and Metrology No. 2832 on December 29, 2018. It establishes the procedure for transferring units from the State primary measurement standard of mass fraction and mass (molar) concentration of water in liquid and firm substances and materials (GET 173) to working standards and measurement instruments.

In the approved verification schedule, the transfer of units of water content from the GET 173 to the measuring units is carried out using reference materials (RMs) [1, 2]. By now there is a limited range of reference materials of approved types for the transfer of units of water content. But use of these RMs is not appropriate, in particular the organization and participation in key, pilot and additional comparisons of national primary measurement standards are required. The reason is matrix effects in most existing RMs. The scope of RMs is limited by strict compliance to regulated procedures, the accuracy of establishing certified values is insufficient, and such RMs have low shelf life.

In this regard, it is relevant to search for chemically stable materials based on crystalline hydrates of alkali and alkaline earth metal salts for the subsequent development of RMs.

The information on the physicochemical properties of crystalline hydrates of salts based on alkali and alkaline earth metals and knowledge of their heating behavior (heat resistance) is fundamental in the development of RM. The crystalline hydrates are the crystals containing water molecules. The formation of these crystals is due to the formation of stronger cationic bonds with water molecules in the crystal lattice in comparison with bonds between cations and anions in an anhydrous substance crystal. Such materials can contain a different number of water molecules - from less than one molecule (for example, calcium sulfate hemihydrate $CaSO_4 \cdot 0.5H_2O$) to twenty water molecules (for example, potassium metatungstate octadecahydrate $K_6H_2W_{12}O_{40} \cdot 18H_2O$), which provides a wide range of measurements of water content.

It is known that, one of the interfering factors in characterizing RM is water adsorbed from air according to the Rebinder classification [3]. At the same time, it should always be taken into account that the stop of the decrease of the mass of the sample during heating does not always reflect the complete removal of moisture from it. Therefore, it is necessary to take into account the effect of the residual water content in establishing of the certified value of RM.

To search for candidate materials for the creation of RM, work was begun to systematize data on the number of water molecules and the temperatures of its removal to an anhydrous state in crystalline hydrates available in various references [4-6]. In order to clarify the literature information, additional studies were carried out using a reference installation based on thermo-gravimetric analysis with mass-spectrometric detection (TGA-MS) as a part of the state primary standard GET 173.

The first RM developed by UNIIM on the basis of crystallohydrate was RM of the mass fraction of water in the dihydrate sodium molybdate (GSO 10911-2017). The range of permissible values of the mass fraction of water is from 14.70 % to 15.00 %, extended uncertainty at k=2 - 0.10 %. The main advantages of GSO 10911-2017 are the constant content of chemically bound (crystallization) water and the absence of matrix effects.

Another promising material for the creation of RM with improved stability characteristics is calcium oxalate monohydrate. The temperature stability of calcium oxalate monohydrate is due to the strength of water bonds with the skeleton of the salt forming the crystallohydrate. Thus, if crystallization water is retained by Van-Der-Waals forces, such substances are stable at temperatures below zero. In calcium oxalate monohydrate, crystallization water is retained by weak intermolecular bonds, so it is easily removed during heating. The low decomposition temperature (200° C) provides the convenience of using of this substance for the calibration of universal water content analyzers.

Thus, by selecting crystallohydrates with sufficient stability and homogeneity and having a different number of water molecules, a "line" of RMs in a wide range of water mass fraction values from 3.0 % to 50.0 % can be created. The results of studies obtained in the selection of materials-candidates for RMs can also be used to create a database of thermograms of crystallohydrates, in the long term issued in the form of a document of the state system of standard reference data (GSSD) [7].

References.

[1] Medvedevskikh M Yu et al. 2019 Creation of transfer measurement standards for implementation of State verification schedule for measuring of water content *Measurement Techniques* **6** 3-10 DOI: 10.32446/0368-1025it.2019-6.

[2] Medvedevskikh S V et al. 2015 General Approaches to the Estimation of Uncertainty in the Results of Reproducing Units of Water Content in Solids and Materials *Measurement Techniques* **58** (8) 926-933

[3] Rebinder P A 1966 Physico-chemical mechanics of dispersed structures (Moscow: Science) 399

[4] Lurie Y Y 1971 Handbook of analytical chemistry 4th ed.; processed and supplemented (Moscow: Chemistry) 456

[5] Knunyants I L 1988-1998 Chemical encyclopedia (in five volumes) [Text] / I. L. Knunyants, N. S. Zefirov (Moscow: Soviet encyclopedia, Great Russian encyclopedia) **1-5**

[6] Ryabchikov D I 1966 Analytical chemistry of rare earth elements and yttrium [Text] / D. I. Ryabchikov, V. A. Ryabukhin (Moscow: Science) 381

[7] GOST R 8.614-2018. State system for ensuring the uniformity of measurements. State service of standard reference data. Basic principles (Moscow: Standartinform) 10

Journal of Physics: Conference Series **1420** (2019) 012007 doi:10.1088/1742-6596/1420/1/012007

Improvement of the reference material planning mechanism for the provision of the necessary nomenclature of various areas of the Russian economy

Egor P Sobina, Sergey V Medvedevskikh, Olga N Kremleva

Ural Research Institute for Metrology (UNIIM), Ekaterinburg, Russia
e-mail: uniim@uniim.ru

Abstract. The paper is about necessity of making organizational decisions, creation of publicly available RM nomenclature planning systems with the involvement of experts, integration into Programs of various level on the measures of state support of developed lists of RMs in Russia. Creation of these lists is necessary to ensure uniformity of measurements in the areas of state regulation in Russia.

The release of the new edition of ISO/IEC 17025 "General requirements for the competence of testing and calibration laboratories", the development of the new standard ISO 17034 "General requirements for the competence of reference material producers", changing laws of countries in the field of ensuring the uniformity for measurements, criteria for accreditation of laboratories, legal entities, individual entrepreneurs, performing work and (or) providing services for ensuring the uniformity of measurements, providers of interlaboratory comparison tests impose a large the degree of relevance on the issues related to reference materials (RMs). The federal executive authorities of the Russian Federation have submitted the issue of planning the RM nomenclature to the state level. That is why, in the Strategy for Ensuring the Uniformity of Measurements of the Russian Federation until 2025, approved by a Government Decree, a significant role is assigned to the issues in the RM area and to the creation, including planning, of the required RM nomenclature.

In the development of the RM planning system in 2018 the Ministry of Industry and Trade of the Russian Federation adopted "Methodical recommendations on planning the nomenclature of reference materials in industries" No.10-751 dated November 29, 2018, providing, *inter alia*, the creation of an Expert Group consisting of specialists from different industry areas. In the framework of the State Reference Material Service (GSSO) UNIIM was entrusted to implement, by the end of 2019, the approaches to planning and regular monitoring of the RM nomenclature provided for by the Methodological recommendations.

These guidelines show the following approaches:

1. Collection and analysis of the initial information recommended for planning the RM nomenclature. The following sources are indicated as sources of initial information:

a) unified single list of measurements related to state regulation;
b) existing technical regulations and a list of standards, the application of which ensures compliance with technical regulations on a voluntary basis;
c) list of Rosstandart orders on the approval of the State verification schemes (SVS) and their annexes, as well as interstate standards, regulating SVS;
d) information on RMs of approved types registered in the Federal Information Fund for Ensuring the Uniformity of Measurements of the Russian Federation (FIF);
e) information on RMs used as verification tools in accordance with the description of measurement instruments of approved types registered in the FIF;
f) information on RMs of approved types used by accredited testing, verification and calibration laboratories when performing measurements (information from documents on laboratory equipment with RMs from the Rosaccreditation Federal State Information System (FSIS);
g) applications and appeals of federal executive bodies, individuals, legal entities and individual entrepreneurs, concerning the need in approved types of RM required for metrological assurance of measurements.

2. Forming the list of RMs reflecting the industries needs, according to the results of initial information analysis, taking into account expert analysis by specialists, appointed to GSSO by federal executive bodies and representing various sectors of the economy.

3. Forming and periodic updating of the Unified Interdepartment Program for the development of RMs of approved types, needed for the metrological assurance of measurements related to government regulation.

For 2019 and the following years, the Strategy defines the work on GSSO development and this question will be in the first place for making organizational decisions, creation of publicly available RM nomenclature planning systems with the involvement of experts, integration into Programs of various level on the measures of state support of developed lists of RMs. Creation of these lists is necessary to ensure uniformity of measurements in the areas of state regulation.

Journal of Physics: Conference Series **1420** (2019) 012008 doi:10.1088/1742-6596/1420/1/012008

The potential of inductively coupled plasma mass spectrometry for measuring the isotopic composition of elements

Olga N Kremleva, Tatyana N Tabatchikova

Ural Research Institute for Metrology (UNIIM), Ekaterinburg, Russia

e-mail: kremleva@uniim.ru

Abstract. Brief information about the development of enriched isotopic composition reference materials (RMs) is presented. The potential of inductively coupled plasma mass spectrometry for measuring the isotopic composition of elements are shown.

Mass spectrometry is a basic or one of the basic (depending on the application) instrumental methods of chemical or biochemical analysis. A variation of the method, namely isotope dilution mass spectrometry, is recognized by the Consultative Committee for Amount of Substance (CCQM) as a method that can be considered the primary measurement method [1].

The primary task in the development of the isotope dilution method in inductively coupled plasma mass spectrometry is the methodological support of the method, including the development and creation of a nomenclature of pure materials (reference materials) of the isotopic composition of the elements.

Ural Research Institute for Metrology conducts research of the possibility of using isotope dilution mass spectrometry, aimed at the development of reference materials of substances and materials and at the same time, systematically develops approved type reference materials of enriched isotopic composition of elements. The research is carried out on an inductively coupled plasma mass spectrometer NexION 300D ("Perkin Elmer"), which is part of the State secondary measurement standard of units of mass fraction and mass (molar) concentration of metals in liquid and solid substances and materials GVET 196-1-2012.

To date, the following reference materials have been developed:

- RM of nickel isotopic composition enriched with ^{60}Ni isotope in nitric acid solution (^{60}Ni RM UNIIM) GSO 10273-2013;
- RM of lead isotopic composition enriched with ^{206}Pb in a nitric acid solution (^{206}Pb RM UNIIM) GSO 10274-2013;
- RM of cadmium isotopic composition enriched with ^{111}Cd isotope, in solution (^{111}Cd RM UNIIM) GSO 10493-2014;
- RM of silver isotopic composition enriched with isotope ^{107}Ag, in solution (^{107}Ag RM UNIIM) GSO 10494-2014.

Considering the significant need for such RMs, materials for RMs of enriched chromium, thallium, magnesium, strontium, rubidium, copper, zinc, etc. isotopic compositions have been prepared for further development.

Characterization of RMs of the isotopic composition of elements is based on the isotopic composition measurement, isotopic ratios. The thermal ionization mass spectrometry method (TIMS) was considered generally accepted for highly-accurate isotope analysis just until the end of the 20th century. However, in recent decades, the inductively coupled plasma mass spectrometry method explosively pressed the TIMS method in isotope analysis. At the same time, the ICP- MS method is quite diverse in instrumentation. These include quadrupole (ICP-Q-MS), time-of-flight (ICP-TOF-MS), magnetic (ICP-SF-MS), and high-resolution devices (ICP-HR-MS) with single-collector and multi-collector ion current detection systems and a number of others.

Most commercial inductively coupled plasma mass spectrometers are quadrupole instruments (ICP-Q-MS). At the same time, the possibility of using quadrupole mass spectrometers with ICP for precision isotopic analysis remains controversial, however, in our opinion, it is quite acceptable for the characterization of RM isotopic composition, intended for the use as a "tracer" for implementing the method of mass spectrometry with isotopic dilution.

The article [2] presents the comparison of the data on the establishment of isotopic ratios $^{87}Sr/^{86}Sr$ in RM of the isotopic composition of strontium carbonate SRM 987 (manufactured by NIST, USA) obtained in different years on various types of inductively coupled plasma mass spectrometers. UNIIM, having this RM available, joined the work and presented additional data on the possibilities of using the inductively coupled plasma quadrupole mass spectrometer (NexION 300D, manufactured by Perkin Elmer, USA) for the determination of isotopic ratios. Obtained results are in good agreement with data from various foreign laboratories.

Thus, it can be assumed that the isotopic composition RM of strontium, enriched with ^{86}Sr, the metrological characteristics of which will be established using the NexION 300D quadrupole mass spectrometer, will not be inferior to its analogs, which have been characterized using traditional isotope methods analysis.

Taking into account the conducted research, it became possible to improve the State primary measurement standard GET 176 in the field of isotope ratios measurements and the use of the mass spectrometry method with isotope dilution.

References.

[1] Minutes from the 5th meeting of the Consultative Committee of the International Parliament on the International Security Council for Europe 1998 (Sèvres, France)

[2] Zoriy M V, Rashad A, Pickhardt C, Mohsen H T, Forstel H, Helal A I, Zahran N F, Becker J S 2003 Routine Method for 87Sr/86Sr Isotope Ratio Measurements in Biological and Geological Samples After Trace Matrix Separation by ICP-MS *Atomic Spectroscopy* **24** (6) 195-200.

Journal of Physics: Conference Series **1420** (2019) 012009 doi:10.1088/1742-6596/1420/1/012009

The research of the possibility to apply the ionchromatographic method to the monitoring metrological characteristics of the working standards of ion activity

Sergey Prokunin, Lidiya Galkina, Vladimir Dobrovolskiy, Narine Oganyan

All-Russian Scientific Research Institute of Physicotechnical and Radio Engineering Measurements (VNIIFTRI), Mendeleevo, Moscow region, Russia, 141570

e-mail: galkina@vniiftri.ru

Abstract. In this work have been researched the possibility to apply the ionchromatography method for the process of controlling and producing of the Working Standards of Ion Activity.

To applying the ionchromatography method for the process of controlling and producing of the Working Standards of Ion Activity was compared two methods: coulometric and ionchromatographic. The aim of the work was to put the ionchromatographic system into process of controlling the production of the Working Standards of Ion Activity (WSIA), which are produced in the VNIIFTRI.

The objectives of this work were:

1. The partly replacement of the coulometric method with the method of ionchromatography in order to unload the state primary standard for indicators of pX ion activity in aqueous solutions [1, 2].
2. The control of the purity chemical reagents and water, which are used for the produce of WSIA;
3. The reduction of the analysis time for quality control WSIA;
4. The inclusion of the ionchromatographic method in the process of monitoring the metrological characteristics of WSIA.

Previously, the WSIA compliance with the specified characteristics was monitored only by a help of the coulometric analysis method, which is quite laborious and time consuming [2, 3]. At the end of 2018, the latest generation ionchromatographic system, which determined the anionic and cationic composition of solutions and water, was commissioned. The advantage of this system is relatively short time needed for the analysis and high accuracy.

At the first stage of our work, according to the instruction manual, the ionchromatographic system was calibrated at three points with using foreign multi-ion calibration standard samples (Figure 1).

Afterwards, the chromatographic analysis of the water used for the production of WSIA was carried out. The evaluation results were positive. No significant concentrations of interfering ions were detected in the presented sample.

At the next stage, the quantitative characteristics were evaluated using the example of K^+ and Cl^- ions in ready-made WSIA samples.

Journal of Physics: Conference Series **1420** (2019) 012009 doi:10.1088/1742-6596/1420/1/012009

The obtained data was used to recalculate the concentrations in the activity of the corresponding ions. Recalculation in activity was performed using the Debye-Huckel equation. Subsequently, the data was compared with the calculated ones. The results are presented in the table 1.

Table 1. Comparison of metrological characteristics of working standards of ion activity

WSIA name	pX theoretical	pX experimental	ΔpX theoretical	ΔpX experimental
WSIA-potassium	1,12	1,12	±0,01	±0,01
WSIA-cloride	1,12	1,12	±0,01	±0,01

The data of the table show the theoretical and experimental values of pX coincide, and the values of the absolute inaccuracy is within the permissible limits.

Summing up the results of this work: was experimental demonstrated that the chromatographic system can be used for control the metrological characteristics and for evaluation the quality of WSIA produced at VNIIFTRI. In the future, we are planned to use method of ionchromatography for developing a preparing procedure of WSIA with main ions content of Ca^{2+}, Mg^{2+}, PO_4^{3-} and others.

References.

[1] Karpov O V 2011 Report to the Federal Agency for Technical Regulation and Metrology. State primary standard of indicators of pX activity of ions in aqueous solutions of Get 171 (Mendeleevo: VNIIFTRI) (In Rus.)

[2] Artemenkov M A , Karpov O V, Kutovoi V D, Zvezdina V A 2010 Coulometric installation as part of a complex of state primary standards *Measurement techniques* (Moscow) **4** 13 (In Rus.)

[3] Karpov O V, Kerimov V A, Lapshin A I 1994 Chemical potentiometry of liquid media. On the metrological support of ionometric, pH-metric and redoxometric measurements *Legal metrology* (Moscow) **1** 23-26

CEN standard "Harmonized counting of atmospheric ultrafine particles" and UFP monitoring initiatives in Europe

Lucia Bustin[1], Torsten Tritscher[1], Jürgen Spielvogel[1], Oliver F. Bischof[1], Jacob Scheckman[2] and Thomas Krinke[1]

[1] TSI GmbH, Neuköllner Str. 4, 52068 Aachen, Germany, email: lucia.bustin@tsi.com
[2] TSI Inc., 500 Cardigan Road, 55126 Shoreview, MN, USA

Abstract. The paper is devoted the standard CEN/TS 16976 described a standardized measuring method with number concentration of UFP. As well as, described TSI 3750-CEN CPC which meet all of these requirements.

1. Introduction

Atmospheric ultrafine particles (UFP) and their impact on human health and our climate are current research topics. UFPs can be present in large numbers in atmosphere, but they contribute little to particulate mass (PM). Some monitoring networks like NABEL (EMPA, 2000) and GUAN (Birmili *et al.*, 2015) use MPSS (Mobility Particle Size Spectrometers) including CPCs (Condensation Particle Counters) as detectors. However, there are also few sites (UBAU) which employ additional CPC for the total particle number concentration to supplement gravimetric PM measurements with real-time measurement of particle number concentration.

It is important to have comparable ambient air monitoring data. While PM2.5 and PM10 measurements are standardized within the EU, this is not the case for the particle number concentration, which is dominated by UFPs.

Comparing UFP number concentrations can be challenging, when CPCs with different D_{P50} detection efficiency diameters, operating conditions, or even working fluids are used. In addition different sampling systems can lead to differences, especially due to particle losses by diffusion.

2. Methods

The European Committee for Standardization (CEN) developed the Technical Specification CEN/TS 16976 as a first step of harmonizing the continuous measurement of particle number concentration in ambient air.

A CPC with a dedicated sampling system (Model 3750-CEN CPC and Model 3772200 CEN sampling system, TSI Inc., Shoreview, USA), both fully compliant with the proposed CEN/TS 16976, have been introduced by Spielvogel et al. (2016). As required, this CPC is a full flow, butanol-based device that counts airborne ultrafine particles from 7 nm (D50) at an aerosol flow rate of 1.0 L/min. It uses an optimized coincidence and dead time correction method to count particles in single count mode up to concentrations of 100,000 particles cm^{-3}. Figure 1 shows the linearity of four TSI CEN CPCs against a reference aerosol electrometer. Other advancements include a pulse height analyser for the CPC that monitors wick health, supersaturation state, and instrument status.

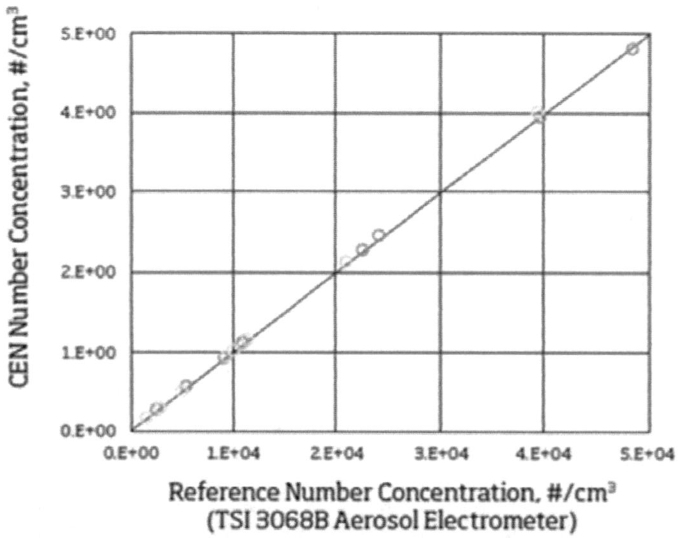

Figure 1. Concentration linearity of CEN-CPC over a wide range. The colored points represent data from four different units tested against an electrometer. The black line shows the 1:1 line.

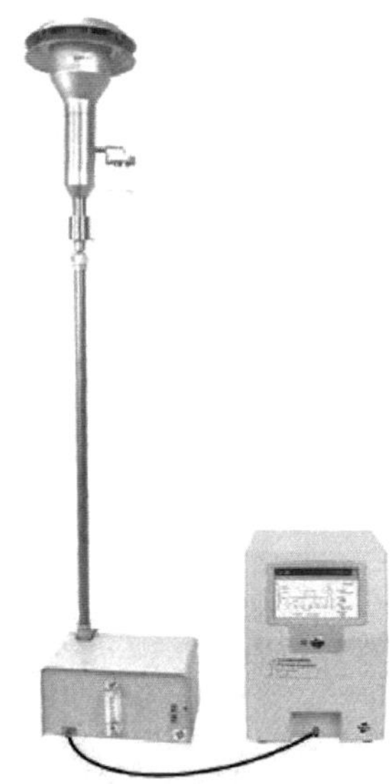

Figure 2. CEN-CPC together with CEN approved Environmental Sampling System (ESS).

3. Conclusions

CEN/TS 16976 describes a standardized method by defining a set of requirements for the CPC, its sampling system, the measurement procedure and the reporting of measurement results that harmonizes the number concentration measurement of fine and ultrafine particles. The 3750-CEN CPC meets all of these requirements and brings the same technology and accuracy to atmospheric monitoring that the Engine Exhaust CPCs (EECPCs) used in type approval testing according to Euro 5b/6 legislations rely on.

Its CEN-sampling inlet systems includes a PM10 sampling inlet, a Nafion dryer and a 3:1 dilution to condition the ambient aerosol if the ambient concentration is too high.

The built-in RH/temperature sensor reports the data to the CPC and all data are recorded in one CEN-conform data file.

4. References.

[1] Birmili W et all 2015 Atmospheric aerosol measurements in the German Ultrafine Aerosol Network (GUAN) *Gef. Reinh. Luft*

[2] CEN/TS 16976:2016: Ambient air, X. Determination of the particle number concentration of atmospheric aerosol (2015)

[3] EMPA: Technischer Bericht zum Nationalen Beobachtungsnetz für Luftfremdstoffe (NABEL), Dübendorf, Switzerland (2000)

[4] Spielvogel J et all 2016 Harmonizing the Measurement of Ultrafine Particles in Atmospheric Aerosol, *Fifth International Conference Nanosafe* (Proceedings – abstract and presentation)

Journal of Physics: Conference Series **1420** (2019) 012011 doi:10.1088/1742-6596/1420/1/012011

CCQM-P194 'The determination of number concentration of colloidal nanoparticles'

D I Belenkii, T M Magomedov, D M Balakhanov

All-Russian Scientific Research Institute of Physical-Technical and Radiotechnical Measurements

Abstract. The paper describes pilot comparisons of P194 'The determination of number concentration of colloidal nanoparticles', measurand, measurement methods and principles. Preliminary results of comparisons and additional measurements of particle sizes and zeta potential of the provided samples are presented. Conclusions are drawn about the possibility of using this sample as a reference material.

1. Introduction

Nanotechnology is actively used to solve scientific and industrial problems. Measurements of the nanoparticles concentration in colloidal suspensions are of great interest for a wide range of industries, including pharmaceuticals, cosmetology, food and microelectronic industries.

However, to date, the methods used to determine the concentration of nanoparticles in liquids are not standardized.

Within the framework of the IAWG, it was decided to carry out the CCQM-P194 comparisons, since this task meets the requirements of the strategic program of pilot comparisons for new and challenging measurands.

During comparisons, participants determined the concentration of gold nanoparticles, according to their chosen method.

2. Measurand

The sample is colloidal spherical gold nanoparticles (stabilized by citrate buffer) with an average diameter of approximately 30 Nm suspended in water. The nominal number of particles is in the range 1-2.5E11 NPs/g. the suspension of nanoparticles (approximately 5 ml) is contained in a sealed ampoule of amber glass, flushed with Ar-gas and sterilized by gamma irradiation. Up to 5 vials of CCQM sample with freeze and 30 + indicators were sent out to all participants. To measure the concentration of gold nanoparticles, the laboratory coordinator recommended the use of dilutions (depending on the measurement method) in a 1mM citrate buffer.

NIST provided one ampoule of PM 8012 (30 Nm gold nanoparticles) to each participant to control sample quality and determine transport efficiency using the spICPMS method. 300 μl sub-aliquots of this sample were sent to each participant. A certificate of analysis for this material with instructions for use is also sent to each participant.

The monodispersity of the sample was tested by the LGC coordinator laboratory.

The concentration of gold nanoparticles in 15 ampoules was measured for 8 hours by particle tracking analysis (PTA, Malvern Panalytical NS500). During one measurement, there were five repeated measurements of the instrument per sub-sample.
Concentration of gold nanoparticles in 15 ampoules was measured for 8 hours by Particle Tracking Analysis (PTA, Malvern Panalytical NS500). During one measurement, there were five replicate instrument measurements per sub-sample.
According to the results of measurements homogeneity uncertainty was 2.49%.
Stability testing by LGC using PTA-samples stored at temperatures up to 60 °C for 1 day and two weeks (fig.1).

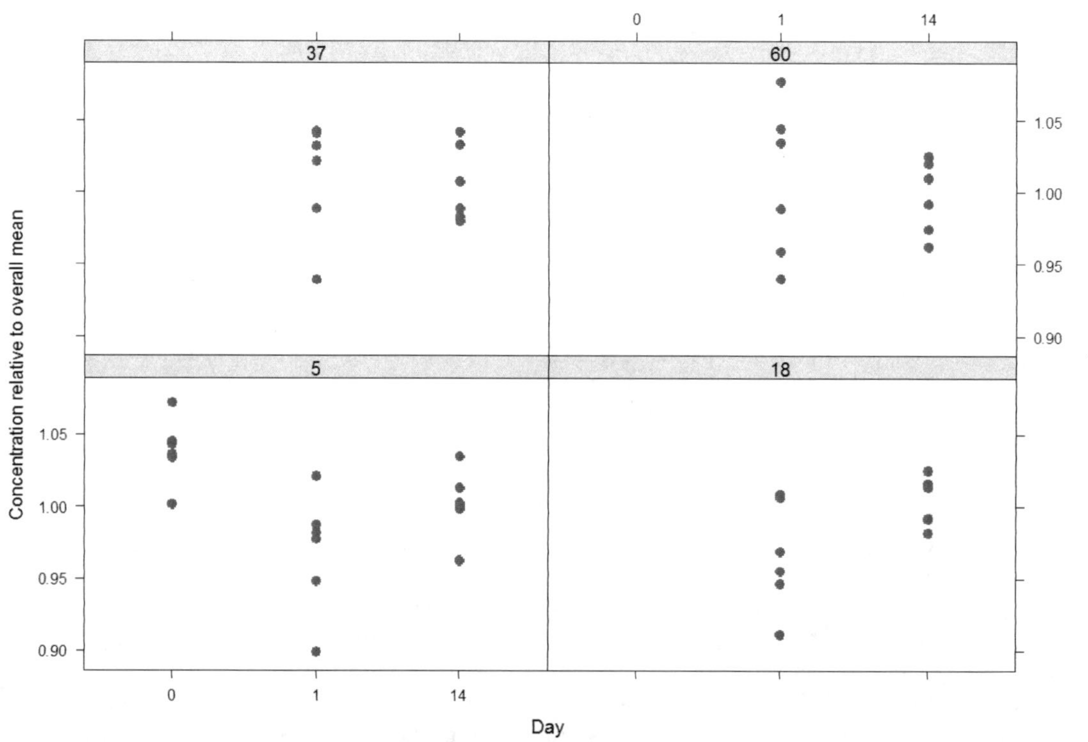

Figure 1. Sample stability tests

3. Measurement methods used by VNIIFTRI

Participants were free to use any suitable method. A full uncertainty budget also had to be included with results. During discussion of spICPMS measurements at the IAWG it was noted that this is a very new method and a key aspect of the pilot study is to gain experience of the measurements and investigate consistency between different laboratories/instruments.. In order to facilitate this, the coordinating laboratory has provided information on aspects of the methodology which have been found to assist reproducible measurements.
For spICPMS measurements, the dwell time recommended for this study depends on the in house capabilities and practice experience of each participant. Overall, dwell time resulting in baseline separation between background signal and particles population are recommended.
Transport efficiency should be determined following the procedure developed by each participant but as specified in PD ISO/TS 19590:2017, but it is recommended to asses transport efficiency at the start, middle and at the end of the measurement batch to be able to assess the within batch variability of this parameter.

During the comparisons we used two methods to measure the concentration of gold nanoparticles:

- The method of dynamic light scattering (DLS);
- Inductively coupled plasma mass spectrometry (sp-ICP-MS).

3.1 Dynamic light scattering

Dynamic light scattering method [1] is used to measure the size of nanoparticles. In this method, the diffusion coefficient of dispersed particles in a liquid is determined by analyzing the characteristic time of fluctuations measured in the intensity of scattered light. The diffusion coefficient is used to calculate the hydrodynamic radius of nanoparticles. Thermal (Brownian) motion of particles causes fluctuations of their local concentration. In turn, these fluctuations lead to local inhomogeneities of the refractive index of the medium. When a laser beam passes through such a medium, part of the light is scattered on these inhomogeneities. The fluctuations of the scattered light intensity correspond to the fluctuations of the local concentration of dispersed particles. Information on the particle diffusion coefficient is contained in the time-dependent correlation function of the intensity fluctuations.

Particle size measurements are made according to the Stokes-Einstein equation (1):

$$d = \frac{kT}{3\pi\eta D} \tag{1}$$

Where:
k = Boltzmann constant;
T =Absolute temperature;
η = Solvent viscosity;
D = Diffusion coefficient.

To calculate the concentration of particles, data on their size, optical properties (imaginary and real parts of the refractive index), polydispersity index and count rate (number of photons per second) are used.

3.2 Inductively coupled plasma mass spectrometry

Inductively coupled plasma mass spectrometry (ICP-MS) [2] is a type of mass spectrometry that uses an inductively coupled plasma to ionize the sample. The sample, an aqueous suspension, is introduced continuously into a standard ICP-MS system that is set to acquire data with a high time resolution (i.e. a short dwell time is used). Following nebulization, a fraction of the nanoparticles enter the plasma where they are atomized and the individual atoms ionized resulting in a cloud of ions. This cloud of ions is sampled by the mass spectrometer and since the ion density in this cloud is high, the signal pulse is high compared to the background signal if a high time resolution is used.

The number of pulses detected per second is a directly proportional to the number concentration of nanoparticles in the aqueous suspension that is being measured. To calculate concentrations, the nebulization efficiency has to be determined first using a reference particle.

Particle concentration measurements are made according to the equation (2):

$$C_p = \frac{C_m}{m_p} \tag{2}$$

Where:
C_p = particle number concentration;
C_m = mass concentration of the particle suspension;
m_p= mass per particle.

Individual particle mass is calculated according to the equation (3):

$$m_p = \frac{I_p t_d}{RF_{ion}} \cdot \frac{V \eta_n}{60} \cdot \frac{M_p}{M_a} \tag{3}$$

Where

m_p = particle mass;
I_p = particle signal intensity in the sample;
RF_{ion} = ICP-MS response for ion standard;
t_d = dwell time;
V = sample flow rate;
η_n = nebulization efficiency;
M_p = molar mass nanoparticle material;
M_a = molar mass analyte measured.

4. Results and discussion

The results of the comparison presented in the preliminary report are shown in figure 2.

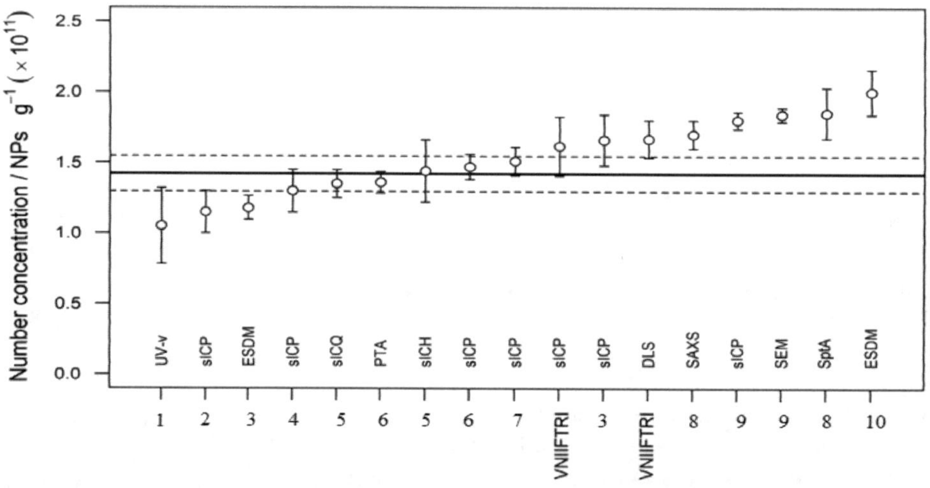

Figure 2. The results of the comparison. Measurement uncertainties are given at k=1, the uncertainty of the median value was 0.124×10^{11} g^{-1}

However, we have carried out additional measurements of the size and zeta potential of the particles, as the values of the concentration varied from vial to vial within 15 %.

According to the results of particle size measurements, the average particle diameter was 30 nm, and the polydispersity reached 21 % (fig. 3).

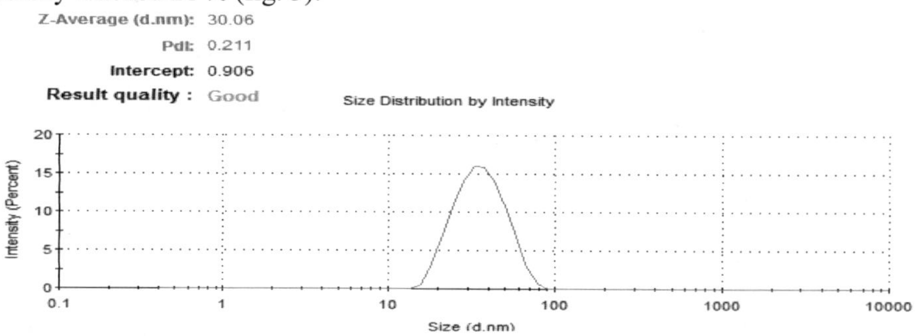

Figure 3. Gold nanoparticles size and polydispersity index

According to the results of measurements of the Zeta potential of the particles, the average value was about 50 mV, and the deviation reached 27 mV (fig.4).

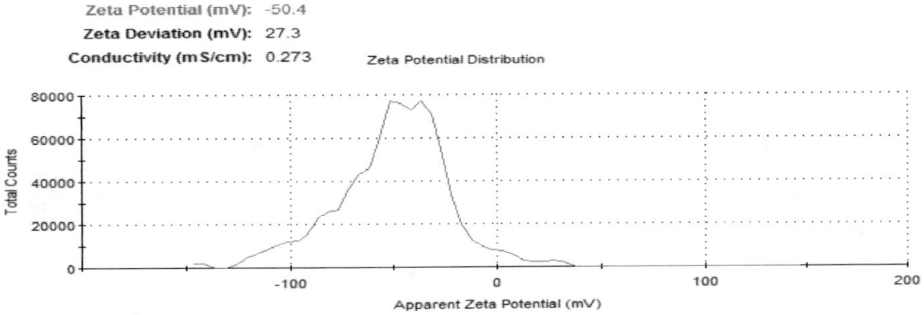

Figure 4. Gold nanoparticles zeta potential

Summing up all the obtained measurement results (size, counting concentration, zeta potential), the use of gold nanoparticle samples provided by the coordinating laboratory as reference material is questionable.

5. References.

[1] ГОСТ Р 8.774-2011 "ГСИ. Дисперсный состав жидких сред. Определение размеров частиц по динамическому рассеянию света"

[2] H E Pace, N J Rogers, C Jarolimek, V A Coleman, C P Higgins and J F Ranville 2011 Determining transport efficiency for the purpose of counting and sizing nanoparticles via single particle inductively coupled plasma mass spectrometry *Anal. Chem.* 9361–9369

Development of reference methods of measurements mass (molar) concentration of proteins in the field of physical and chemical analysis of blood

N Y Griazskikh, A V Ivanov, N P Muravskaya

All-Russian Scientific Research Institute of Optical and Physical Measurements (VNIIOFI), Moscow, st. Ozernaya, 46, Russia, e-mail: vniiofi@vniiofi.ru

Abstract. The work is devoted to the development of reference methods for measuring the mass concentration of proteins in blood plasma that would meet modern metrological requirements (high accuracy of reference values, metrological traceability, etc.), which will allow the use of modern high-tech methods of physicochemical analysis.

Physico-chemical properties of blood include: blood density (specific gravity), rheological properties (viscosity, erythrocyte suspension stability (ESS)), acid-base balance (ABB), colloid osmotic blood pressure.

The physical and chemical properties of blood are greatly influenced by the protein content (concentration). For example, the relative density of blood plasma is mainly determined by the concentration of proteins and is 1.029 - 1.032. With injuries, infections, and many acute diseases, acute phase proteins appear in the blood for several hours, which increases the erythrocyte sedimentation rate (ESR). Proteins are involved in the acid-base reactions of the body, so the protein buffer system (protein-COOH / protein-COONa) is the main intracellular buffer. Proteins are amphoteric compounds and can neutralize both acids and alkalis. Oncotic pressure depends on the amount of protein [1, 2].

Thus, the accuracy of measuring the concentration of various proteins plays an important role in the diagnosis, treatment prescription.

According to GOST ISO 17511-2011 "Medical devices for in vitro diagnostics. Measurement of values in biological samples. Metrological traceability of values attributed by the calibrator and control materials." The ideal end point of the traceability chain is to determine the appropriate unit of the International System of Units (SU), however, the choice of stages and levels at which traceability ends for a given value depends on the availability of measurement procedures and higher-level standards. In many cases, traceability of measurements cannot currently be carried out to a level of a higher order than the technique chosen by the manufacturer or the manufacturer's working calibrator. In such cases, correctness refers to the level of the calibration hierarchy at which the internationally recognized reference measurement procedure and / or reference material (calibrator) become available.

According to the Joint Committee on Traceability in Laboratory Medicine (JCTLM), the main reference method for determining proteins is the method of immunonephelometry and immunoturbidimetry [3, 4].

Table 1. JCTLM Reference Methods.

Analyte	**Метод**
Albumen	Optimized immunoturbidimetry / immunonephelometry
a1-anti cymotrypsin	Optimized immunoturbidimetry / immunonephelometry
a1-anti trypsin	Optimized immunoturbidimetry / immunonephelometry
Ceruloplasmin	Optimized immunoturbidimetry / immunonephelometry
Complement C3	Optimized immunoturbidimetry / immunonephelometry
Complement C4	Optimized immunoturbidimetry / immunonephelometry
Glycoprotein	Optimized immunoturbidimetry / immunonephelometry
Haptoglobin	Optimized immunoturbidimetry / immunonephelometry
Glycohemoglobin HbA1c	Cation exchange HPLC; LC-MS
Hemoglobin	Optimized immunoturbidimetry / immunonephelometry
Immunoglobulin G	Optimized immunoturbidimetry / immunonephelometry
Immunoglobulin M	Optimized immunoturbidimetry / immunonephelometry
a2-macroglobulin	Optimized immunoturbidimetry / immunonephelometry
Total protein	Biuret reaction
C-reactive protein	Optimized immunoturbidimetry / immunonephelometry
Transferrin	Optimized immunoturbidimetry / immunonephelometry
Transteritin	Optimized immunoturbidimetry / immunonephelometry

VNIIOFI has carried out work on the development and approval of a primary reference method for measuring the mass concentration of immunoglobulin G in blood serum, on the technique registered in JCTLM [4].

As a result, the primary reference method for measuring the mass concentration of immunoglobulin G in blood serum by the method of immunoturbidimetry / immunonephelometry No. ФР.ПР1.31.2019.00003 was approved.

Table 2. Metrological characteristics of the primary reference measurement procedure

Range of measurements of mass concentration of IgG, g/dm^3	2,00 - 30,00
Type A standart uncertainty, u_A, %	2,00
Type B standard uncertainty, u_B, %	1,59
Total Standard Uncertainty, u_C, %	2,55
Extended Uncertainty at coverage ratio k = 2, U, %	5,11

Currently, VNIIOFI is working on the development of reference methods for measuring other proteins (albumin, immunoglobulin A, etc.), as well as on the development of reference materials of various types of proteins.

References.

[1] Alipov A N, Muravnik L M, Ronzhina N L, Safyannikov N M 2010 Medical laboratory photometric devices and complexes St. Petersburg *Renome* p 504
[2] Menshikov V V 1987 Laboratory research methods in the clinic. Directory Moscow p 368
[3] https://www.bipm.org/jctlm
[4] Blirup-Jensen S 2001 Protein standardization III: method optimization. Basic principles for quantitative determination of human serum proteins on automated instruments based on turbidimetry or nephelometry, *Clin. Chem. Lab. Med.* **39** (11) p 1098-1109

Special measures of radionuclides activity to metrological support of measurements in radiation control

V Dyachkova, M Zotova and N Tonkih

All-Russian Scientific Research Institute of Physicotechnical and Radio Engineering Measurements (VNIIFTRI), Mendeleevo, Moscow region, Russia, 141570

Abstract. The possibilities of FSUE VNIIFTRI for production and calibrating special measures of activity (imitants of radioactive contamination of environmental objects) are presented. A description of the measures of activity with different geometries and materials for the calibration of spectrometric, radiometric and aerosol equipment is given. The importance and practical importance of the production of these imitants is determined by the fact that they are widely used for metrological support of radiometric and spectrometric measuring instruments because it make it possible to calibrate the measuring instruments for conditions that are closest to the real ones, thereby reducing the error of the measurement results.

1. Introduction

Currently, radiation and environmental safety is one of the priority areas. Therefore, there is a vast task to ensure the unity and reliability of measurements of radionuclide activity of environmental objects in industry.

One of the main activities of the Research Department of Metrology of Ionizing Radiation (NIO-4) of VNIIFTRI is to develop and produce various imitants (measures of activity) of radioactive contamination of environmental objects, nuclear and building materials, agricultural products and food. The imitants are both monolithic and granular bulk samples with variable parameters.

The imitants are widely used for metrological support of radiometric and spectrometric measuring instruments since it possible to use them for calibration the measuring instruments in actual-use environment, thereby reduced the error of the measurement results [1].

The special measures of activity of radionuclides are produced both like an approved type of the measures and for the specific needs of the Customers [2].

The measures of activity are radioactive substance with a specific measurement geometry (a specific design and relative position of the object of study and the detector of the measuring instrument). The form of the measures of activity excludes pollution of environmental and equipment when used them in actual-use environment. They can be divided into four groups which are described following.

2. Description and possibilities of production of the special measures of activity

The first group is point measures of activity. In this case, the active part is made of an aliquot of nuclide solution localized application with known specific activity on a substrate with following sealing. The point measures of activity are shown in figure 1.

Figure 1. The point measures of activity

The second group is surface measures of activity. In this case, the active part is made of a nuclide solution uniform distribution with known specific activity on the surface of a substrate with following sealing. The surface measures of activity are shown in figure 2.

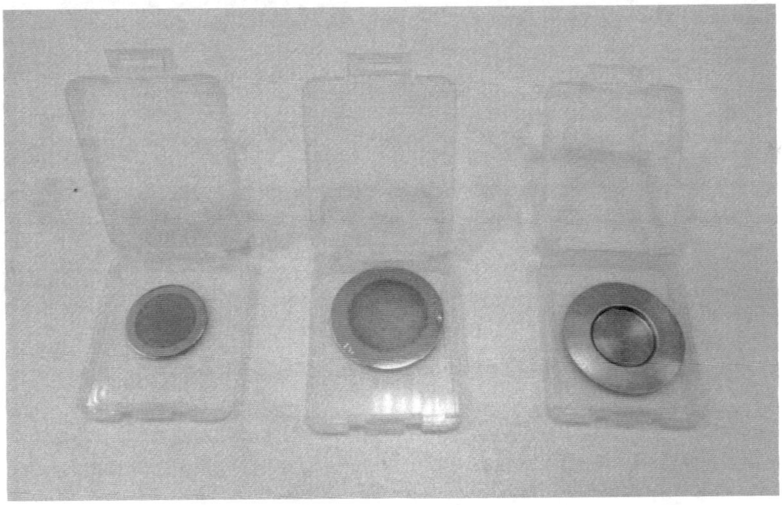

Figure 2. Surface measures of activity

The third group is solid-state volumetric measures of activity, the active part of which is made of a nuclide uniform distribution with known activity in the volume of sources, followed by sealing. If necessary, the volume of the source can be made in metal or organic forms. The source can also be made in a temporary form followed removal. The solid-state volumetric measures of activity are shown in figure 3.

Figure 3. The solid-state volumetric measures of activity

The fourth group is bulk granular volumetric measures of activity consisted of sealed granules with radionuclide or mixture of uniformly distributed radionuclides. The bulk granular volumetric measures of activity are shown in figure 4.

Figure 4. The bulk granular volumetric measures of activity

Almost any material characterized varuios natural environment objects (plant material, sand, soil, water, gas) can be used in the production of the radioactive contamination imitants. In the most cases, basis is an epoxy resin with known activity, and density of the source determines the degree of equivalence to real samples. For the bulk and the solid-state imitants (imitants of soil, sand and food samples), the density can vary between 0,5 – 2,4 g/cm3, for imitants of water, air and gas 0,05 – 1,0 g/cm3. The list of radionuclides, which can be used as a basis of the measures of activity, are presented in table 1 [3]. The list is not full.

Journal of Physics: Conference Series **1420** (2019) 012013 doi:10.1088/1742-6596/1420/1/012013

Table 1. The list of radionuclides, which can be used as a basis of the measures of activity

Radionuclide	Half Life, yrs	Radionuclide	Half Life, yrs
^{207}**Bi**	32,9	^{232}**Th**	$1{,}4 \cdot 10^{10}$
^{133}**Ba**	10,51	^{40}**K**	$1{,}248 \cdot 10^{9}$
^{137}**Cs**	30,17	^{90}**Sr**	28,79
^{44}**Ti**	60,0	^{36}**Cl**	$3{,}1 \cdot 10^{5}$
^{22}**Na**	2,6	^{239}**Pu**	$2{,}411 \cdot 10^{4}$
^{241}**Am**	432,6	^{60}**Co**	5,2713
^{226}**Ra**	1600	^{152}**Eu**	13,537

The main characteristics of the sources are given in the table 2 [2].

Table 2. The main characteristics of the sources

Name of characteristic	Characteristic value
Energy range, keV	4 – 6130
Activity (specific activity) range, Bq (Bq/kg, Bq/l)	от 10 до $1 \cdot 10^{6}$
Expanded uncertainty at coverage factor k = 2 (corresponding to a confidence interval of approximately 95 % assuming a normal distribution), %	± (5 – 20)
Unevenness distribution of activity on the surface (by volume),%, no more	5
Average technical life time from commissioning, years	from 5 to 20
Operating conditions:	
- the temperature range, °C	from - 50 to + 50
- the relative humidity range at the temperature 30 °C, %	to 95
- atmospheric pressure, kPa	from 25 to 105

There are the necessary radiochemical and technological equipment for the production of measures of activity at VNIIFTRI.

Metrology supporting of spectrometric measuring instruments with the special activity measures imitated real environmental samples radionuclides pollution allows decrease the error of the measurment instruments. This is also minimazing the risks of false desision in conformity assessments due to measurement results in field of radiation technologies: radiation safety at emergency situations, control of agricultural products, environmental objects (soil, yes, air), medicine (for example, radiation therapy in oncology), etc.

3. References.

[1] Sarkisov E R 2005 Standards of the unit of activity of radionuclides of VNIIFTRI and metrological support of radiation ecology *Collection of scientific works of VNIIFTRI* **52 (144)** 74

[2] Radionuclide sources of photon radiation of metrological purpose, closed, IMN-G (*Electronic resource)* https://fgis.gost.ru/fundmetrology/registry/4/items/355459

[3] MGFK.412128.001TU 2010 Radionuclide sources of photon radiation of metrological purpose, closed, IMN-G *Technical conditions* (Moscow: Russian Metrological Institute of Technical Physics and Radio Ingineering).

Dosimetric system on the basis of EPR-spectroscopy with use state primary standards of power of the absorbed dose and Standard of number of the paramagnetic centers

Leskov A S , Kuvykina M B, Tenishev V P

All-Russian Scientific Research Institute of Physicotechnical and Radio Engineering Measurements (VNIIFTRI), Mendeleevo, Moscow region, Russia, 141570

e-mail: tenishev@vniiftri.ru

Abstract. This paper presents an overview of the current state of ensuring the measurement uniformity of absorbed dose of ionizing radiation during the treatment of food and agricultural products in Russia. The way of measuring the absorbed dose of ionizing radiation in the range of absorbed doses up to 1 kGy with Primary State Standard GET209-2014 and GET83-2017 is proposed.

Application of ionizing radiation for food and agricultural products treatment [1] requires new and precise ways of controlling the effects of radiation on the substance and ensuring the quality of products without its deterioration as a result of ionizing radiation. The main tool of such control is measurement of the absorbed dose (AD) in a product and precise definition minimum and maximum of allowed AD values directly after radiation treatment. The correct and precise measurement of AD provides the proper solution for such technological tasks as sterilization of medical products, maintaining quality of food products at their radiation processing, radiation resistance of materials, etc.

Table 1 shows the ranges of absorbed doses for irradiation of food and agricultural products (from world practice).

Table 2 shows serial measuring tools of absorbed dose.

In this work the way of storage, transfer of dose rate unit and AD unit and measurement of AD using the Primary State Standard of absorbed dose rate (GET209-2014) [2] and the primary state standard of the number of paramagnetic centers GET209-2014 in the field of radiation technologies is considered. The main idea of this method is using the EPR-spectrometry with application of reference measures of number of the paramagnetic centers:

- the radiation of EPR sensitive material on GET209-2014 with strictly defined doses in the required range of the absorbed doses;
- the measurement of the EPR signal of the irradiated samples on EPR spectrometer, which is calibrated with standard of the number of the paramagnetic centers [3].

Journal of Physics: Conference Series **1420** (2019) 012014 doi:10.1088/1742-6596/1420/1/012014

Table 1. The ranges of absorbed doses for irradiation of food and agricultural products

Purpose of radiation treatment	Dose range, kGy		Dmax/Dmin
Delay growing of bulbs and tubers	0.03–0.15	0,03 – 0,1	3 - 5
Insect disinfection and	0,15 -0,5	0,03 – 0,2	3 - 6
Destruction of parasites	0,25 -1,0		4
Improving and ripening of fruits	0.03–0.15		5
Prevention of infection by insects, parasites	0.07–1.00	1 – 1.3	14 – 1,3
Delayed rotting and spoiling meat	1.50–3.00	2,0 – 7,0	2 – 3,5
Extended shelf life (meat, berries, mushrooms, etc.)	1,0 – 3,30		3,3
Reducing the risk of infection with pathogens		1.00–7.00	7
Increasing sanitary conditions for spices	до 10.00		3
Sterilization of packaged meat	25 - 70		
Sterilization of meat, seafood, poultry, etc.		30- 50	1,6
Disinfection, decontamination of individual additives and ingredients		10 - 50	5

Table 2. Serial measuring tools of absorbed dose

Detector	Range, kGy	Type of radiation	Uncertainty, %
Film (standard)	5 – 50	γ, β, e-	3 – 7
RM PD(F)E-5/50			
GSO 7865-2000			
Film (standard)	5 – 50	γ, β, e-	7 – 12
RM PD(F)R-5/50			
GSO 7865-2000			
Film (working)	30 - 200	γ, β, e-	7 – 15
RM PD(F)R -30/200			
GSO 7903-2001			
Film (working)	1 – 10	γ, β, e-	7 – 15
RM PD(E)-1/10			
GSO 8916-2007			
Block (working)	0,05 – 0,6	γ	3 – 7
RM PD (DTS)-0,05/10	1 –10		7 – 15
GSO 9447-2009			

The absorbed dose (D) is proportional to number of paramagnetic centers in the sample and can be defined from the equation:

$$D \sim k \cdot N = k \cdot \frac{A_D \cdot (\Delta H_D)^2}{A_{ref} \cdot (\Delta H_{ref})^2} \cdot N_{ref}$$

where,

N – the number of the paramagnetic centers in the irradiated sample;
A_D – the EPR line amplitude of the irradiated sample;
A_{ref} – the EPR line amplitude of the reference sample;
N_{ref} – the number of the paramagnetic centers in the reference;
ΔH_D, ΔH_{ref} – the width of the EPR lines of the irradiated and reference samples respectively, mT;
k – the proportionality coefficient defined from calibration function.

The number of the paramagnetic centers and thereby the value of the absorbed dose are defined by amplitude and the width of the EPR line as the irradiated sample and the reference sample. Therefore, different materials will show a different EPR-response to the same absorbed dose. This makes the calibration of EPR-spectrometers necessary for providing traceability to the state standards.

There are the results of researches of different materials for the broad range of the absorbed doses in this work.

References.

[1] Donchenko et al. Metrological support of scientific developments and industrial technologies with the use of intensive sources of ionizing radiation 2019 *II International Scientific Forum Nuclear science and technologies* (Almaty, Republic of Kazakhstan) (In Russian)

[2] Generalova V V et al. 2015 State primary special standard of absorbed dose rate unit of intensive photon, electron and beta radiation for radiation technologies GET 209-2014 *Almanac of Modern Metrology* **5** 54-74 (In Russian)

[3] Leskov A S State primary standard for the number of paramagnetic centers: yesterday, today, tomorrow *Almanac of Modern Metrology* **1** 2019. № 1 (17). 35-43 (In Russian)

Journal of Physics: Conference Series **1420** (2019) 012015 doi:10.1088/1742-6596/1420/1/012015

The development of the standards of the EPR-spectrum using the quartz monocrystal the morion

Kuvykina M B, Trotsenko N P, Leskov A S

All-Russian Scientific Research Institute of Physicotechnical and Radio Engineering Measurements (VNIIFTRI), Mendeleevo, Moscow region, Russia, 141570

e-mail: kmb@vniiftri.ru

Abstract. The results of the development of the standard of the EPR-spectrum are represented. This standard is used for determination of EPR-spectrometer resolution. This standard can be used as a work standard, providing traceability to the State primary standard.

The Electronic Paramagnetic Resonance (EPR) is a phenomenon of resonance absorption of electromagnetic energy in the centimeter or millimeter wavelength range of the paramagnetic substances placed in a constant magnetic field.

The EPR is used in a research of the paramagnetic centers and their environment in material. EPR-spectrometers allow to determine the concentration and identify paramagnetic particles in any aggregate state [1, 2].

Evaluation of sensitivity and resolution of EPR-spectrometers is possible with using the standards of the number of paramagnetic centers and EPR spectrum standard (ESS).

Traditionally, a perylene solution in sulphuric acid is used to determine the resolution of EPR-spectrometers. However, the capillary with solution of perylene is extremely sensitive to influences of external factors of the environment and is unstable over time. As a result its EPR-signal level decreases irreversible to zero. Thus its application for measuring of ESS is possible only for short time periods [3].

Minerals (natural crystals) can be a good alternative for organic solutions with insufficient resistance to environmental conditions. They have narrow and stable EPR lines. The monocrystal of natural quartz the Morion seems to be the most promising [4, 5].

It was found that the EPR-spectrum of the Morion quartz crystal samples is anisotropic. It has strong dependence on the orientation of the crystal in magnetic field of the EPR-spectrometer.

The narrow lines periodically appear in the EPR spectrum while rotating the quartz Morion sample with random orientation around a rotation axis. Their width is less than 6 mkT, characterizing super-thin structure. However, it is extremely difficult to reproduce the results even at the same sample.

We have an idea how to receive the identical samples from a quartz crystal the Morion with the EPR-spectrum containing necessary narrow lines. The idea was to create crystals with known axis orientation.

The orientation of the quartz the Morion crystal was performed with X-ray goniometer 1H-1C.

Measuring samples were located in the resonator and EPR-spectra at various rotational angles of crystal axes were recorded.

These research showed the reproducible dependence of the EPR-spectrum of quartz from sample to sample from orientation of axes in magnetic field of the EPR-resonator.

It was found that when aligning crystal axes to the direction of magnetic field in the resonator superposition of many lines of the EPR-spectrum (the so-called not resolved thin structure) was observed. The EPR-spectrum example with the coincidence of the magnetic field direction with Z crystal axis presented in figure 1.

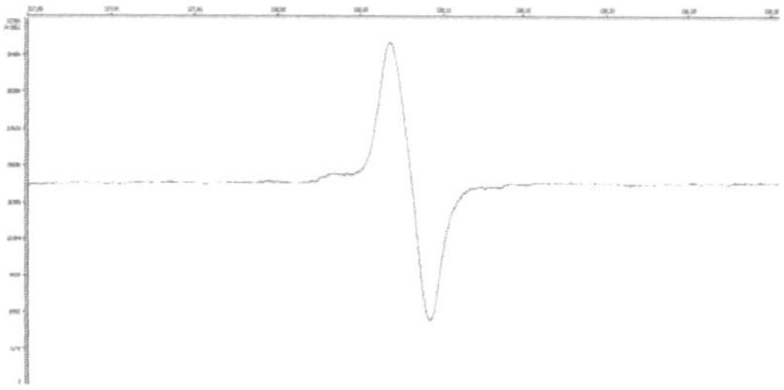

Figure 1. A quartz EPR-spectrum with the coincidence of the magnetic field direction with Z crystal axis.

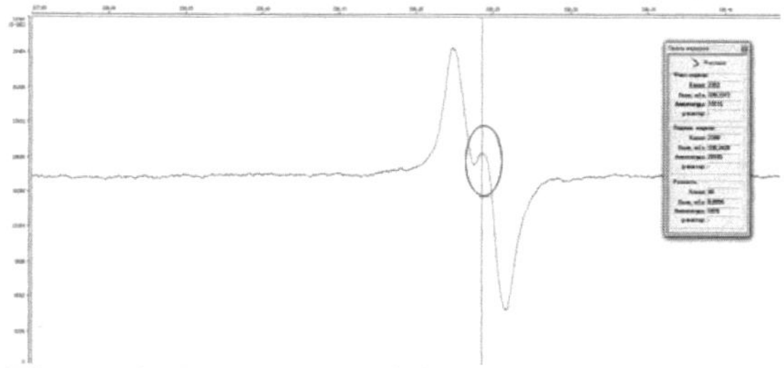

Figure 2. Emergence of the narrow line in a quartz EPR-range the Morion at a deviation of the direction of axis Z of a crystal from the direction of magnetic field in the resonator.

Similar spectra are received on all oriented samples.

It was also found that with turn axes of the crystal at a small angle (less than one degree), the hyperfine structure of a EPR-spectrum appears (figure 2).

The line width of hyperfine structure of the EPR-spectrum is less than 6 mkT and steadily reproduced from measurement to measurement. For ESS samples based on the quartz the Morion it was established that the line width of hyperfine structure of EPR-spectrum is 4-6 mkT. That is fully compliant with the requirements for measuring the ESS standards.

Thus, the standard of ESS has been developed. It reproduces EPR-spectrum and uses the quartz the Morion with the known values of linewidth of EPR-spectrum and the hyperfine structure component of the EPR-spectrum. This standard can be used for resolution determination and calibration of values of magnetic field of EPR-spectrometers.

References.

[1] Blumenfeld L A, Tikhonov A N 1997 Electronic paramagnetic resonance *Soros educational journal* **9** (In Rus.)

[2] Ingram D 1961 Electronic paramagnetic resonance in free radicals (Moscow) (In Rus.)

[3] Leskov A S, Kuvykina M B 2019 Standard spectrum measures for EPR spectrometry *Metrology in the XXI century. Materials of the VI scientific-practical conference of young scientists, postgraduates and specialists* (VNIIFTRI) 172-174 (In Rus.)

[4] Shakurov G S 2105 High-frequency EPR spectroscopy of impurity paramagnetic ions in dielectric and semiconductor crystals (Kazan, Doct. Thes.) (In Rus.)

[5] Aminov L K, Kurkin I N Malkin B Z 2013 Superfine structure of EPR spectra and optical spectra of impurity f-ions in dielectric crystals *Solid state physics* **55** ed.7 (In Rus.)

Questions of metrological support for bioanalysis. Catalytic activity of enzymes

E V Kulyabina[1], T V Kulyabina[2], V V Morozova[3], O N Melkova[1], E A Guskova[1]

[1]VNIIMS, Moscow, Russia, e-mail: kuliabina@vniims.ru
[2]Independent consultant, St. Petersburg, Russia, e-mail: fdsay@mail.ru
[3]Student, Far Eastern Federal University, Vladivostok, Russia, e-mail: lerapuri@gmail.com

Abstract. The relevance of metrological support of measurements carried out in laboratory medicine is examined to increase the accuracy of analyzes results, namely, measurements of the catalytic concentration of the human body enzymes. The development of reference materials with a long shelf life is discussed. The results of the development of Russian reference materials of the catalytic concentration of α-amylase and alkaline phosphatase are discussed. Based on the measuring model (Beer-Lambert-Bouguer law), sources of uncertainty are established. The developed reference materials make it possible to measure the catalytic concentration of enzymes with high reliability, make the correct diagnoses at the early stages of the disease, and make it possible to achieve the world level of metrological support of laboratory medicine in the field of measuring enzyme parameters.

What does modern man begin with? The modern man begin with health, with intelligence, with the ability to properly organize time, with incredible efficiency and determination, as well as with a burning desire to do something for the existence of humanity every day. Of course, this is just our point of view, but this statement looks reasonable, and the probability of challenging this thesis seems small.

So, health comes first. There are individuals who were lucky enough to be born with good hereditary traits to the correct functioning of the body systems (estimated by no more than 4% of the total population of the Earth [1]), their genome allows, without worrying much about health, nevertheless, living healthy for a long time. However, there are other individuals, and there are many of them (the remaining 96%) who need to pay attention to the functioning parameters of their body - cholesterol, blood pressure, blood and urine sugar levels, enzyme activity. The accuracy of the analysis results of these parameters directly depends on the development of the metrological support of Bioanalysis, laboratory medicine.

The development of biotechnology in the real world is at an incredible pace. For accurate measurements, unprecedented opportunities of the latest discoveries and achievements of science and technology are used. The importance of studying the catalytic processes occurring in the human body was noted by the great biochemist Albert L. Lehninger ‘For the existence of life, two basic conditions must be met. First, a living organism must have the ability to reproduce itself. Secondly, the body must be able to effectively and selectively catalyze chemical reactions‘[2].

How important is the study of catalytic processes, including those occurring in the body, and the fact that in the history of the Nobel Prize, which in chemistry was awarded 166 times, 73 times of them were awarded for discoveries made in bioanalysis, and for discoveries directly related to the study of catalysis and catalysts, 18 times.

Currently, millions of medical tests are performed annually in the world. Data on Nobel Prizes awarded, as well as information on the number of bioanalytical measurements carried out annually in the world, give only a rough presentation of the relevance of obtaining accurate measurements results of quantities related to bioanalysis. Accurate measurements in bioanalysis make it possible to conduct early diagnosis of diseases, which helps prevent the uncontrolled course of pathology and cure a person of a serious disease; control the quality of pharmaceuticals; provide accurate results of analyzes of the content of toxic, psychoactive, consciousness-altering agents in a patient, determine the possible presence of doping in blood, urine samples, etc.

At the moment, accurate and operational methods of analysis and measuring instruments are being developed, improved and created, which make it possible to solve these pressing tasks. Large foreign and domestic enterprises are create such measuring instruments is carry out.

Issues of ensuring metrological traceability of the chemical and biological measurements results to physical quantities primary standards, to primary reference procedures are regularly discussed and resolved at meetings of the Consultative Committee on the Quantity of Substances (CCQM) of the International Bureau of Weights and Measures (BIPM).

Twice a year, meetings of three working groups related to bioanalysis are held - the protein analysis working group (PAWG), the cell analysis working group (CAWG), and the nucleic acid analysis working group (NAWG).

Russia in this type of measurement is traditionally represented by VNIIM named aft. D I Mendeleev, and from the spring of 2015, VNIIMS joined them. VNIIM has 5 primary measurement standards in the field of physicochemical measurements; VNIIMS is currently creating the primary reference procedure for measuring the catalytic concentration of the α-amylase.

Recall that the metrological support system for any type of measurement is based on the availability of relevant regulatory legal acts, the state primary standard or primary reference measurement procedure, the state verification scheme for transferring unit size to subordinate measuring instruments or a hierarchical scheme, reference materials and unit size transfer methods.

International symposia on reference materials BERM 14, which took place in October, 2015 in Washington, USA, and BERM 15, which took place in September, 2018 in Berlin, Germany, were devoted to the creation and use of biological reference materials and reference materials for ensuring traceability of measurements in the field of environmental protection. At BERM 14, 23 companies were represented - the world's leading manufacturers of reference materials, 99 oral presentations and 108 poster presentations were presented. Many countries are closely engaged in solving the problem of creating stable, with long shelf life reference materials of biological substances. Attracted interest a report on a very original 'biological eggs'.

'Biological eggs' (BioBall^{TM}, Biomérieux) are reference materials (SRM) enclosed in an egg-shaped shell consisting mainly of calcium (Figure 1).

Figure 1. 'Biological egg' [3].

Due to its tightness, a constant microclimate is maintained inside this shell, which allows biological elements (DNA, proteins, and peptides) enclosed in it to maintain their properties almost unchanged. The percentage of SRM stored in BioBall that remains stable for 12 months is approaching 100 % (97 %). While the same biological substances, traditionally stored in glass containers, remain stable in about 70 % of cases (according to Biomérieux [3]).

In Russia, primary measurement standards are created and improved, verification schemes are developed, methods are developed, but reference materials of approved types that play one of the most important roles in bioanalysis (for the purpose of transferring the size of a unit) are currently practically absent. With only an approximate assessment of the need for SRM, calculated in thousands, about 30 were registered with the Federal Information Fund for Ensuring the Uniformity of Measurements.

Our institute conducts extensive research and development of Russian reference materials used for the diagnosis of diseases, namely reference materials of catalytic activity / concentration of enzymes. The first in the ever-expanding list of sought-after enzymes were α-amylase and alkaline phosphatase, reference materials of which were created as part of the experimental design work of 'Bioanalysis'.

Given that measurements of the catalytic concentration are carried out by spectrophotometric method, the measurement model is based on the Beer-Lambert-Bouguer law, and the equation of optical density measurements is converted to the form (1):

$$A = \varepsilon l c = \varepsilon l \frac{c_0}{D} \tag{1}$$

where:

A - the optical density of the solution;
ε - the coefficient of molar extinction;
l - the length of the optical path (thickness of the cell);
c - the concentration of 4-NP in the reaction solution *;
c_0 - the concentration of 4-NP before dilution *;
D - the dilution factor.

* - in the case of α-amylase

Differentiating equation (1) we obtain equation (2):

$$\frac{dA}{dt} = \frac{d(\varepsilon l \frac{c_0}{D})}{dt} = \frac{\varepsilon l}{D}\frac{dc}{dt} \tag{2}$$

The reaction rate can be calculated from equation (3):

$$r = \frac{dc}{dt} = \frac{D}{\varepsilon l}\frac{dA}{dt} = \frac{31k}{1012 \times 0.01} \times 1000 = 3063k \quad \mu\text{kat/L} \tag{3}$$

where:

$r = \frac{dc}{dt}$ - α-amylase activity
D - dilution factor
ε - the coefficient of molar extinction;
l - the length of the optical path (thickness of the cell);
c - the concentration of 4-NP in the reaction solution;
$k = \frac{dA}{dt}$ - the slope of the regression.

Based on the measuring model (equation 1), the following sources of uncertainty are accepted for consideration:

- dilution factor - takes into account the uncertainties of all volumetric utensils used for the preparation of solutions;
- the length of the optical path - depends on the accuracy of the measured thickness of the cuvette, is taken from the certificate for the cuvette or determined empirically;
- the slope of the direct regression - depends on two factors - the first - on the measurements of optical density (taken from the certificate for the spectrophotometer) and time, the second - on the linearity of the regression;
- correlation of the slope angle, dilution factor and optical path length;
- deviation of the measurement method;
- pH of solutions;
- solutions temperature.

The created reference materials have the characteristics given in table 1.

Table 1. Metrological characteristics of SRM [4]

№	Name of Characteristics	Dimension	Value of the characteristic	Limit deviation - extended uncertainty of catalytic concentration measurements at (k = 2)
SRM α-amylase				
1	Catalytic concentration of the main component	kat/dm^3	$5{,}0 \cdot 10^{-7} \div 1 \cdot 10^{-5}$	0,5 ÷ 7 %
SRM alkaline phosphatase				
1	Catalytic concentration of the main component	kat/dm^3	$5{,}0 \cdot 10^{-7} \div 1 \cdot 10^{-5}$	0,5 ÷ 8 %

The shelf life of reference materials is 12 months.

Figure 2 shows the appearance of reference materials of α-amylase and alkaline phosphatase.

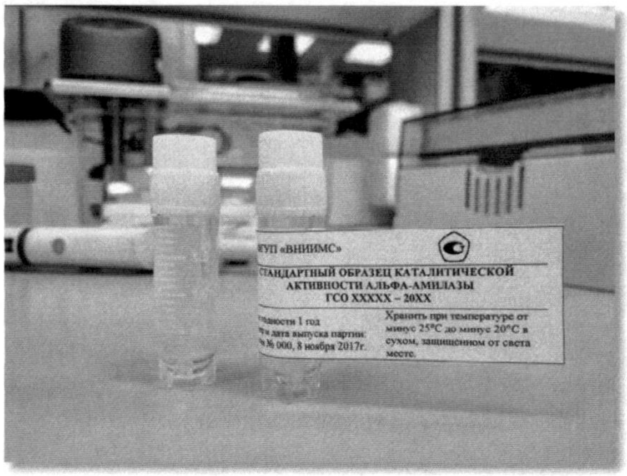

Figure 2. Appearance of reference materials of α-amylase and alkaline phosphatase

As a result, conditions have been created for import substitution and import advances, conditions have been created for Russia to enter the forefront in the manufacture of affordable high-tech, competitive metrological support tools - reference materials for laboratory medicine; national reference materials of catalytic concentration of the analyzed enzymes of world-class were created.

The developed reference materials were created taking into account the existing world experience and the level of technological development using the results and procedures of international comparisons CCQM BIPM. The studies performed make it possible to measure the catalytic concentration of the identified analyzed substances with high reliability and to make the correct diagnoses at the early stages of the disease, and allow achieving the world level of metrological support of laboratory medicine in the field of measuring enzyme parameters.

Thus, the work on the creation of methods and means of metrological support of bioanalysis brings us closer to more accurate test results and the achievement of a healthy person.

References

[1] Nicholas J Kassebaum, Amelia Bertozzi-Villa, Megan S Coggeshall at all 2014 Global, regional, and national levels of maternal mortality during 1990–2013: a systematic analysis of the Global Burden of Disease study 2013 *Lancet* **384** 980–1004 DOI: 10.1016/S0140-6736(14)60696-6

[2] D Nelson, M Cox 2011 Basics of Leninger Biochemistry: in 3 t. (Moscow: Bi-nom. Laboratory of Knowledge) **1** 269

[3] *https://www.biomerieux-usa.com/industry/bioball*

[4] R & D Report Development of reference materials of biological catalytically active substances for the calibration / calibration of biological measurement and reference methods for measuring the composition (purity) of the original pharmaceutical substances 2019 *the Bioanalyst code* **3.2** p 22

Journal of Physics: Conference Series **1420** (2019) 012017 doi:10.1088/1742-6596/1420/1/012017

Results of research of transfer measurement standards in the form of pure substances

Pavel V Migal

UNIIM, 4 Krasnoarmeyskaya Str. Ekaterinburg, uniim@uniim.ru

Abstract. The paper presents the summarized results of the development of transfer measurement standards in the form of pure substances and the methods of transferring units, characterizing chemical composition from transfer measurement standards to routine measurements with the use of reference materials. The work was performed in the framework of the research project "Purity" and experimental project "Purity2b".

Activity of international metrological organizations, such as Consultative Committee for Amount of Substance of International Committee of Weights and Measures (CCQM CIPM), COOMET and national metrology institutes is aimed at ensuring the uniformity of measurements and traceability of measurement results to state measurement standards of units. The main and very important part of the system of traceability of measurement results are materials with a certified value of component content and established uncertainties of these values (reference materials of an approved type), for creation of which pure substances, certified at a higher reference level, are needed, i.e. transfer measurement standards.

Transfer measurement standards in this sense are solid pure inorganic substances with an established value of the mass fraction of the main component with the highest accuracy or with accuracy, sufficient to transfer a unit from the state primary measurement standard to the next links of traceability chain. Pure substances are widely used as information carriers in quantitative chemical analysis [1, 2, 3], e.g. as solutions, used for calibration of composition analyzers [4]. These pure substances are studied in accordance with the procedures for reproducing units, which characterize the content of components using state primary and secondary measurement standards and are part of the state primary measurement standards.

The development of reference measurement standards was carried out in the framework of research project under the code "Purity" (2015-2016) and experimental project under the code"Purity2b" (2017-2019).

The paper demonstrates the principal possibility of ensuring metrological traceability when transferring units from reference measurement standards in the form of pure substances to composition reference materials, intended for calibration of analytical equipment both in metrological tasks and for routine measurements (Figure 1).

Approaches, related to the evaluation of metrological characteristics of transfer measurement standards in the form of pure inorganic substances, are harmonized and implemented by approval of the document MI 3560-2016 "Recommendation. GSI. Evaluation of the measurement uncertainty of mass fraction of the main component in inorganic substances". "Data base regulations. "Transfer measurement standards in the form of high-purity substances", which establishes general requirements for transfer measurement standards, their development, purpose and use, is developed.

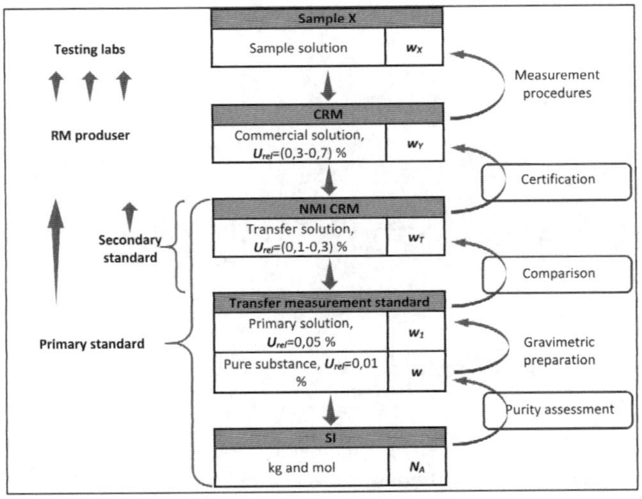

Figure 1 Metrological traceability chain for elemental analysis. It shows the link of mass fraction of w_x element in unknown sample X to SI (International system of units) NA is Avogadro constant; U_{rel} is the relative expanded uncertainty (k=2; P=0,95) [5].

References.

[1] V B Baranovskaya 2016 The synergistic effect of method combination in analytical chemistry of high-purity substances and recyclable metal-containing raw materials:thes. cand. chem. sciences *Giredmet* (Moscow) 330

[2] Jochen Vogl et.all 2016 CCQM-P149 Purity determination of zinc to be used as primary standard for zinc determination (Berlin)

[3] H Kipphardt, R Matschat, J Vogl, T Gusarova et.all 2010 Purity determination as needed for the realisation of primary standards for elemental determination - status of international comparability *Accreditation and Quality Assurance* **15** 29-37

[4] S V Medvedevskikh, E P Sobina, P V Migal et all 2014 On the use of pure inorganic substances in analytival measurement metrology *Reference materials* **3** 58-67

[5] H Kipphardt, R Matschat, O Rienitz, D Schiel, W Gernand и D Oeter 2006 Traceability system for elemental analysis *Accred. Qual. Assur.* **10** 633-639

X-ray absorption spectral analysis of heavy metals with Si-PIN detectors

S M Osadchii[1], A A Petukhov[1] and V B Dunin[2]

[1]All-Russian Scientific Research Institute of Physicotechnical and Radio Engineering Measurements (VNIIFTRI), Mendeleevo, Moscow region, Russia, 141570, osm@vniiftri.ru
[2]Joint Institute for Nuclear Research, Dubna, Moscow oblast, 141980 Russia

Abstract. An X-ray absorption spectroscopy method for the detection of heavy metals at the *K*-absorption edges is implemented using the designed Si-PIN detector. The limits of detection of heavy metals and uncertainty of measurement are estimated by the results of measurements of the thickness of gold and lead foil.

1. Introduction

In many technological operations, such as baggage check, ore dressing, sorting of scrap or waste electronic products, there is a need to detect certain chemical elements. Despite the abundance of tools and methods for elemental analysis, there are very few real devices that can be built into existing conveyor systems and possess the necessary performance to detect certain chemical elements in objects on a conveyer. X-ray Absorption spectroscopy analysis (XASA), which uses the attenuation of the flow of X-ray quanta at the *K*-absorption edges. In this work, the possibility of using the XASA method with the designed Si-PIN detector for the express determination of heavy metals in conveyor systems is considered and the resolution of the method is estimated. To determine heavy metals by the XASA method, a device with a detector possessing the necessary response rate when the analyzed material moves along a conveyor, operating at room temperature without additional cooling, and having sufficient resolution and an upper limit of the operating range of about 100 keV is required. Due to the fact that silicon detectors are widely used in X-ray scanners and the technology of their manufacture is well-established, this study considers the possibility of using Si-PIN detectors for the XASA of heavy metals at lines of the *K* series in a special configuration when X-ray radiation is directed along a layer, i.e., along the long side of the detector. The peculiarity of using Si-PIN detectors is that the analysis was carried out in the range covering the characteristic lines of the *K*-series of heavy metals. To do this, the detector is oriented so that the radiation is directed along the PIN structure, i.e. along the long side of the detector. The detector used in the measurements was developed for X-ray diffraction analysis [1] and then modified for use in XASA in the framework of this study. Detector dimensions are as follows: width 1.5 mm, length 12 mm, and thickness of silicon wafer 0.5 mm. The spatial resolution in this geometry, when the radiation is directed along the layer, is quite acceptable for scanners and can be about 0.5-1.5 mm. The lack of forced cooling eliminates problems with evacuating the detector and substantially simplifies its design. The resolution rating of the detector gives an energy resolution of 1.65 Kev on the 59.5 Kev line of the ^{241}Am source in non-forced cooling mode at room temperature of 22 °C. Detailed specifications of the detector are given in [2].

In this study, the efficiency of the XASA method using the developed detector is experimentally evaluated. In the considered cases, a small amount of a heavy-metal element is located behind a screen made of a metal with a lower atomic number. For example, a microcircuit with gold-plated contacts is mounted in a device case with steel walls or a device electronic board made with the use of copper contacts coated with lead solder is mounted into an aluminum case. For these cases, the detection limits are experimentally determined and the error in measuring the thickness of the gold coating and the lead layer is estimated.

2. Experimental setup

The scheme of the experimental setup is shown in figure 1. The silicon detector is oriented with the end face toward the received radiation. The area of the end of the Si-PIN detector is 0.75 mm^2. A RAPAN X-ray radiation source and an X-ray tube with a tungsten anode were used in the setup. The voltage in the tube was U = 150 kV. The detector was located at a distance of 820 mm from the X-ray tube, so that an extended object could be placed in the illuminated volume, as in a typical X-ray scanner window with a size of 60 x 80 cm. The intensity of tube radiation is very high on tungsten K-lines and at lower values of energy, which causes a large detector load. To eliminate this problem, a filter made of materials with small atomic numbers was used to correct the spectrum and ensure normal operation of the detector, which was located behind the collimator of the tube. The detector was placed in a lead case with a wall thickness of 2 mm. The thickness of the lead collimator of the detector was 10 mm, and the aperture diameter in the collimator was 3.5 mm.

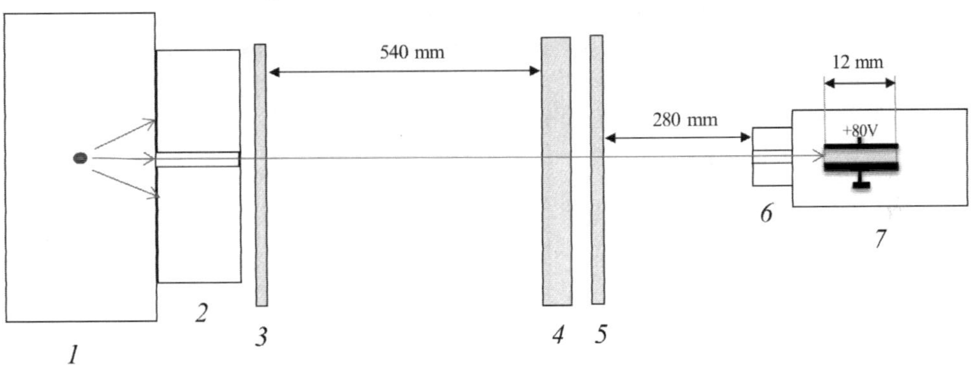

Figure 1. Scheme of the experiment: *1* is an X-ray tube; *2* is a lead collimator; *3* is a low energy filter; *4* is a screen; *5* is a sample; *6* is a lead collimator of the detector; *7* is a detector in a lead case.

The relationship between the incident radiation intensity I_0, transmitted radiation intensity I, and coordinate x along the beam-propagation direction is determined by the following formula:

$$x\,\mu_{h,l} = -\ln(I/I_0)_{h,l} \tag{1}$$

where $\mu_{h,l}$ is the linear absorption coefficient, and the h and l indices refer to the energy windows above and below the absorption edge, respectively.

The x coordinate in Eq. (1) is measured in length units, and μ is measured in inverse length units; $x\mu$ can be expressed as $x\rho(\mu/\rho_t)$, where ρ_t is the reference value of the density and ρ is the density of the used material. In the limit of narrow windows near the absorption edge, the following formula is obtained:

$$x_a = [\ln(I_h/I_l)]/[\mu_l(s-1)] \quad (2)$$

where s is the absorption jump.

Formula (2) assumes a linear dependence of x_a on the real thickness l_m, measured mechanically. Factors affecting linearity and measurement error x_a are thin absorption edge structure, detector noise and collimator fluorescence.

In case of wide windows analyzed in [3], in addition to the above factors affecting the linearity and measurement error, the dependence of the detector sensitivity coefficient on the energy and the unevenness of the X-ray tube radiation spectrum will be manifested.

3. Measuring the thickness of metal films by the XASA method

The effectiveness of the proposed method for analyzing heavy metals was experimentally verified using a setup whose diagram is shown in figure 1. Experiments were conducted, in which the element to be analyzed was lead and the material behind which it is located was a 6-mm-thick steel sheet in one case, and 16-mm-thick aluminum and 1.2-mm-thick copper sheets in the other case. Figure 2 illustrates the comparison of the lead-foil thicknesses measured by the XASA method and with a micrometer gauge (l_m).

Since the sample under study consisted of several layers of technical foil, the exact calculation of the absolute thicknesses is impeded by a lack of accurate values of the density. Therefore, calibration by means of a reference sample with the maximum thickness was performed after measurements. The B type uncertainty was assumed zero for this thickness.

The value linearly depends on the value in the range from 0.02 to 0.4 mm. When the sample thickness is more than 2 mm, the dependence becomes nonlinear. This is because the intensity I decreases and the noise level of the detector begins to have an effect. The detection limits of an element depend on the measurement error. If calibration and measurements were carried out with the same steel screen, the measurement error is about 5 μm. If screens made of different materials are used (a steel screen for calibration, and copper and aluminum screens for measurements), then the assessment gives a resolution of 18 μm in the case of measuring the lead-foil thickness. Replacing the screen material (steel, copper, and aluminum) affects both the change in the intensity of the X-ray quanta flux from the X-ray tube and the change in the spectrum. Both factors lead to an increase in the measurement error.

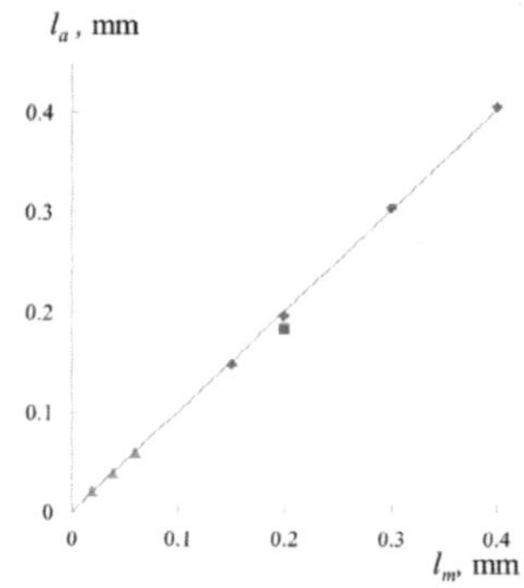

Figure 2. Alignment of the lead-foil thicknesses measured by the XASA method (l_a) and with a micrometer gauge (l_m): rhombs correspond to the first series of measurements with a 6-mm-thick steel screen; triangles correspond to the second series of a 6-mm-thick steel screen; squares correspond to the measurements of a 0.2-mm-thick lead foil behind a two-layer screen comprised of 1.2-mm-thick copper and 16-mm-thick aluminum sheets; $l_a = x_a k$, where k is the calibration coefficient determined from the equality $l_a = l_m$ at $l_m = 0.4$ mm.

Figure 3. Absorption spectra near the absorption edge (E = 80.725 keV) of gold foils with thicknesses of (*1*) 12, (*2*) 24, and (*3*) 36 μm. The energy windows selected (*4*) below the absorption edge and (*5*) above the absorption edge; N is the spectrometer-channel number; Δn is the number of counts in the channel minus the average number of counts over all channels.

Experiments on measurements of the thickness by the XASA method were carried out, in which gold foil samples with thicknesses of 6 to 36 μm were placed behind a steel screen with a thickness of 6 mm. Figure 3 shows the deviations of the intensity from the average value near the absorption edge for various gold-foil thicknesses.

For comparison, the gold-foil thicknesses measured by the XASA method and with a micrometer gauge are shown in figure 4. In contrast to the previous experiment, the spectrum was measured 10 times, and ten thickness values and then their mean, as well as the A and B type uncertainties, were calculated. The measurement uncertainties calculated for each series is given in Table 1. In further calculations, an expanded uncertainty averaged over six series was used, which is 2.7 μm. Energy windows with a width of about 3.5 keV were chosen, and the interval between them was about 3 keV.

Table 1. Uncertainty in measuring the gold-foil thickness by the XASA method

Thickness, μm	6	12	18	24	30	36
Uncertainty of type **A**, μm	1.5	2.3	3.2	2.1	3.1	2.4
Uncertainty of type **B**, μm	1.1	2.0	1.1	0.1	0.8	0
Total standard uncertainty, μm	1.9	3.1	3.4	2.1	3.2	2.4

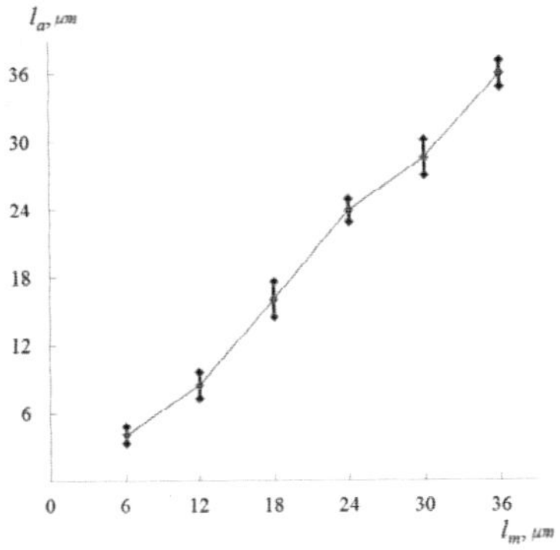

Figure 4 Alignment of the gold-foil thicknesses measured by the XASA method (*la*) and with a micrometer gauge (*lm*). The vertical lines show the A type uncertainty in ten measurements; $l_a = x_a k$, where k is the calibration coefficient determined from the equality $l_a = l_m$ at l_m = 36 μm.

Figure 4 demonstrates that the dependence of the value on the value is close to a linear dependence in the range from 0.006 to 0.036 mm under the selected experimental conditions. With a distance of 820 mm between the X-ray tube and the detector, the relative standard uncertainty of the weight measurement, i.e., the ratio of the minimum weight of the sample per unit area to the weight of the shield per unit area, is 0.1%; the minimum weight of the detected metal gold is 0.15 mg. Similar estimates for the heavy metal thallium show that such indicators are sufficient to detect microdoses from 0.1 mg of thallium in a steel cell with a wall thickness of up to 3 mm.

To simplify the spectrum-analyzer scheme and reduce the acquisition time, a two-channel analyzer with fixed windows can be implemented using the example of the detector described in [1]. By changing the threshold levels of the detector, one can retune the analyzer for the required metal. By expanding the windows to the level of currently used scintillators, it is possible to obtain a distribution over atomic number groups and an object image that are standard for scanners.

Additional ways to reduce the measurement time are as follows: an increase in the current of the X-ray tube, transition to a pulsed mode of synchronous detection, and a decrease in the distance between the X-ray tube and the detector.

4. Conclusion

The performed studies show the possibility of using the ARSA method with Si-PIN detector in sorters, X-ray scanners and inspection equipment to search for objects containing heavy metals, as well as for the enrichment of ores containing heavy metals and the determination of precious metals in waste products of electronic technology.

One of the proposed options for scanners: at first, all objects are analyzed in the normal mode, and then selected objects with large atomic numbers, which are darkened in the picture, are analyzed further by scanners with the option of controlling certain elements.

To extend the functionality, two detectors tuned to different metals can be used in the same scanner.

References.

[1] Osadchii S M, Petukhov A A and Dunin V B 2019 *J. Surf. Invest.: X-Ray, Synchrotron Neutron Tech* **13** **4** 690-694

[2] Osadchii S M, Petukhov A A and Dunin V B 2019 *Measurement Techniques* **5** 65-68

[3] Osadchii S M, Petukhov A A and Dunin V B 2019 *J. Surf. Invest.: X-Ray, Synchrotron Neutron Tech* **13** **4** 683-689

Metrological support of ion-selective electrodes

Sergey Prokunin, Vladimir Dobrovolskyi, Lidiya Galkina, Daria Vengina

All-Russian Scientific Research Institute of Physicotechnical and Radio Engineering Measurements (VNIIFTRI), Mendeleevo, Moscow region, Russia, 141570

e-mail: galkina@vniiftri.ru

Abstract. In this paper the determination of calcium ions concentration in drinking water using ion-selective electrodes are presented

Recently, due to the development of new types of ion-selective electrodes, the ionometric method of analysis is increasingly used. Compared to other methods of analysis, the ionometric method with ion-selective electrodes has several advantages. Ionometric measurements are carried out in a short time, and herewith the usage of sophisticated measuring equipment is not required. Therefore, these electrodes are widely used in research the rate of passage chemical reactions and in the regulation of technological processes at various enterprises of the chemical industry.

It is also important that measurements using ion-selective electrodes belong to the group of non-destructive testing.

If ion-selective microelectrodes are being used, it becomes possible to measure liquid samples with a volume of up to 1 milliliter. At the same time, it should be noted that preliminary sample preparation is not required. Measurements may be carried out in turbid solutions and even in gels; long-term operations of purification, distillation and extraction of the sample are excluded. Therefore, measuring systems based on measurements of electrode potentials are simple to put into operation. Ionometry is also widely used to determine various biologically active compounds and medical supplies.

Another area of applying ionometry is hydrochemical researches, where it is necessary to determine the concentration of cations (sodium, potassium, calcium and magnesium) and anions (chloride, fluoride, nitrate, sulfate and carbonate) in various types of waters: surface, sea and rain.

The expansion of the scope ionometric method has led to the need to create constant conditions for verification (calibration) of ion-selective electrodes. For metrological support of this measurement area in Russia, the State primary standard of indicators of pX activity of ions in aqueous solutions of GET 171-2011 is used, which today has the highest metrological characteristics in Russia.

The aim of our work is to develop a stable mono-element solution of a divalent calcium ion, which is necessary for determining ion-selective electrodes.

In this paper the results of determining the concentration of calcium ions in drinking water using ion-selective electrodes are presented.

During the development process, a pilot batch of calcium-based mono-cell solution samples was prepared, ion-selective electrodes were also calibrated, and the concentration of calcium ions in various samples of drinking water was measured.

Calcium chloride was used to make the monoelement solution. Calcium chloride was subjected to further purification and dehydration. After then, a chemical analysis was performed on content of the main component. A solution of divalent calcium ion was prepared in dionized water at a temperature of t = 20 with a specific conductivity of $5 \cdot 10^{-4}$ $Ohm^{-1} \cdot m^{-1}$. The prepared solution was stored in a tightly closed plastic container in a dark place, to prevent photochemical destruction.

The Nernst equation is used in all analytical methods with ion-selective electrodes. In order to establish the accurate adequacy of work the electrochemical circuit of Nernst equation, ion-selective electrodes are checked. Calibration curves are used to determine the activities and concentrations of ions in the analyzed samples.

Measurements of the electromotive force in the electrochemical cell were carried out on the installation including the following elements: potentiometer (1), ion-selective electrode (2), reference electrode (3), measured sample of calcium solution (4), thermostat (5), thermometer (6), a mixing device (7). The installation is shown on Figure 1.

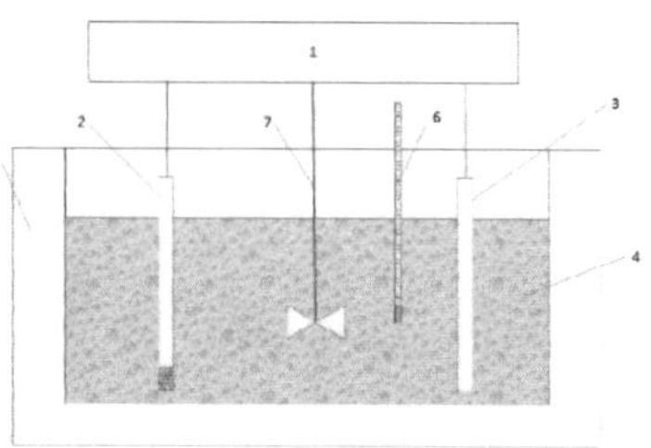

Figure1. Installation for measuring the electromotive force of an electrochemical cell.

A high-precision pH meter was used as a potentiometer with a voltage measurement error of 0.1 mV. The ion-selective electrode (for calcium) (2) was subjected to conditioning before work. Conditioning the electrode involves soaking it in a solution of calcium chloride with a concentration corresponding approximately in the middle of linear portion on the calibration curve. A silver chloride electrode of the ESO-01 type was used as a reference electrode (3). All measurements were carried out at a temperature of t = 20 with using a thermostat (5). Analyzed solutions were mixed with using an electric stirrer to reduce response time of the electrode (7).

Standard solutions of calcium chloride with calcium ion concentrations of 0.001, 0.01, 0.1, 1, 10, 30, 100, 300, 1000, 4000, 10000, 40000, 92000, mg/dm^3 were used for construct the calibration graph. Auxiliary solution of 1 M potassium chloride was used for maintain a constant ionic strength.

The measurements were carried out in order of increasing concentration of calcium ions in solution. After each measurement, the electrodes were washed with distilled water and wiped with filter paper. An each standard solutions in volume 40 cm^3 and an auxiliary solution in volume 10 cm^3 were successively poured into 50 cm^3 measuring cylinders and mixed. Then the cylinders were immersed in the thermostat and thermostated, after that equilibrium potentials were measured.

Measurement results are presented on Figure 2.

As can be seen on the Figure 2, the calibration curve has a linear section in the range of Ca^{2+} concentrations from 10^0 to 10^4 mg/dm^3. Slope of the linear section corresponds to Nernst and equal 28.9 (2.3RT/zF = 58.1/z) mV/z. Detection limit of the Ca^{2+} ion was found at intersection of the asymptotes linear sections is 0.33 mg/dm^3. Deviation from linearity at high concentrations is associated with an abnormally high coefficient activity of $CaCl_2$ salt in this concentration area [5].

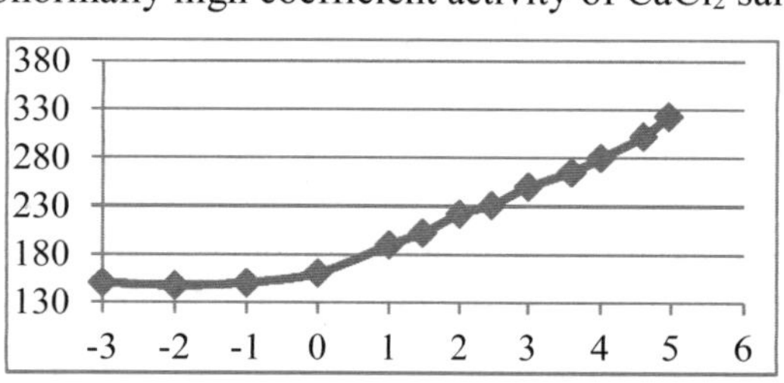

Figure 2. Calibration curve for the calcium selective electrode (standard calcium chloride solutions)

Attempts were made to use $CaNO_3$ salt as the standard titer for determination of Ca^{2+} ion. On picture 2 of the dashed line, shows the calibration curve for calcium nitrate. The narrower linear section from 10^0 to 10^2 mg/dm^3 has a slope different from the Nernst 25,0 (2.3RT/zF = 29) mV/z. Such an electrode function of a calcium selective electrode is associated with the influence of the NO_3^- anion on the operation electrode membrane. Calcium nitrate can partially dissolve in the organic phase of membrane. Thus, it was concluded that calcium nitrate cannot be used as a standard titer for calibration of a calcium selective electrode.

The paper also presents the results of determining concentration of calcium ions in drinking water with using ion-selective electrodes.

Measurement of the electromotive force of the analyzed samples was carried out on the installation, shown in picture 1.

The experimental procedure was as follows: in a measuring cylinder in volume 50 cm^3 were added to the analyzed sample in volume 40 cm^3 and of the auxiliary solution in volume 10 cm^3. It was mixed, thermostated and the value of the equilibrium potential was measured. After each measurement, the electrodes were washed with deionized water and wiped with filter paper. Two parallel measurements were carried out for each water sample. The logC value was found and the concentration of calcium ions was determined from the calibration graph. The accuracy of the determination corresponds to the error of slope the linear portion calibration curve was found. The inaccuracy is constant throughout the linearity range. The measurement results are shown in table 1.

Table 1. The content of calcium ions in various samples of drinking water, measured using ion-selective electrodes.

Samples of drinking water	Concentrations of Calcium ions, mg/dm^3
Mendeleevo settlement:	
- tap water	71,2±6,4
- spring 1	95,1±8,6
- spring 2	95,3±8,6
Zelenograd city:	
- tap water 4 microdistrict	44,0±4,0
- tap water 5 microdistrict	56,1±5,1

It should be noted that the measurements uncertainty by ion-selective electrodes was 9%, which is less than the permissible error for the measurement of drinking water hardness [6].

During the development process, an experimental batch of mono-element calcium solutions was produced. Ion-selective electrodes were calibrated, and the concentrations of calcium ions in various sources of drinking water was measured.

References.

[1] Nikolskiy V P 1937 *Journal of Physical Chemistry* **10** 495 (In Rus.)
[2] Petukhin O M, Zhukov A F 1979 Ionometry (Moscow: MChTI) (In Rus.)
[3] Kumman K 1980 Work with ion-selective electrodes tr.from Germ. (Moscow: Mir)
[4] Koryta I, Shtulik K. 1989 Ion-selective electrods (Moscow: Mir) (In Rus.)
[5] Chemist's Handbook 1964 3 Chemical equilibrium and kinetics, properties of solutions, electrode processes (Moscow)
[6] GOST 27384-2002 Water. Norms of measurement error of indicators of composition and properties (with Changes 1)

Journal of Physics: Conference Series **1420** (2019) 012020 doi:10.1088/1742-6596/1420/1/012020

Development of a conductivity sensor for measuring conductivity of ultrapure water

A M Smirnov, I G Chernikov

D I Mendeleyev Institute for Metrology (VNIIM), e-mail: A.M.Smirnov@vniim.ru

Abstract. The work is devoted to the research results of the developed piston conductivity cell, equipped with additional «guard» electrodes for equipotential protection. Cell able to work both in flow and in static measurement mode.

In 2019, the research work "Development of an experimental sample of a conductivity cell, providing measurements of conductivity in the range from 10^{-5} to 10^{-3} S/m", was completed. The relevance of this work is due, on the one hand, to the development of energy and pharmacology in the Russian Federation, in which put forward stringent requirements for water purity. On the other hand, there is a technical and metrological lag of the Russian Federation in the field of measuring the conductivity of highly purified water, so the range of the state primary conductivity standard of liquids is limited to 10^{-3} S/m.

The demand of this work is also confirmed by supplementary comparison Euramet.QM-S12 "Electrolytic conductivity at pure water level" (2018-2019). In total, this allows to name the issue of expanding the range of the state primary conductivity standard of liquids one of the most actual issue in the field of electrochemical measurements.

In the developed cell, the ability to change the cell's constant by a value measured with the required accuracy is implemented, which allows you to control the effects occurring at the border of the electrode and the solution and affecting on the measurement of conductivity. Since when changing the distance between the electrodes, only the intrinsic resistance of the liquid volume changes, and the influence of the near-electrode processes remains unchanged.

The cell consists of two coaxial capacitances formed by external (large) and internal (small) cylinders. A small cylinder forms a measuring volume intended to investigate liquid. In the static measurement mode the thermostatic liquid is pumped through the external volume which is between the large and small cylinders.

The “generator” electrode moves vertically, and its plane is stayed parallel to the plane of the measuring and “guard” electrodes. The movement of the electrode is controlled by a micrometer.

The "guard" electrode has the same potential with the measuring electrode and designed to reduce the influence of distortion of the current at the edges of the electrodes on the measurement result. And the distribution of current density in the volume of the solution enclosed between the electrodes is uniform.

The model of the developed cell is shown in Figure 1.

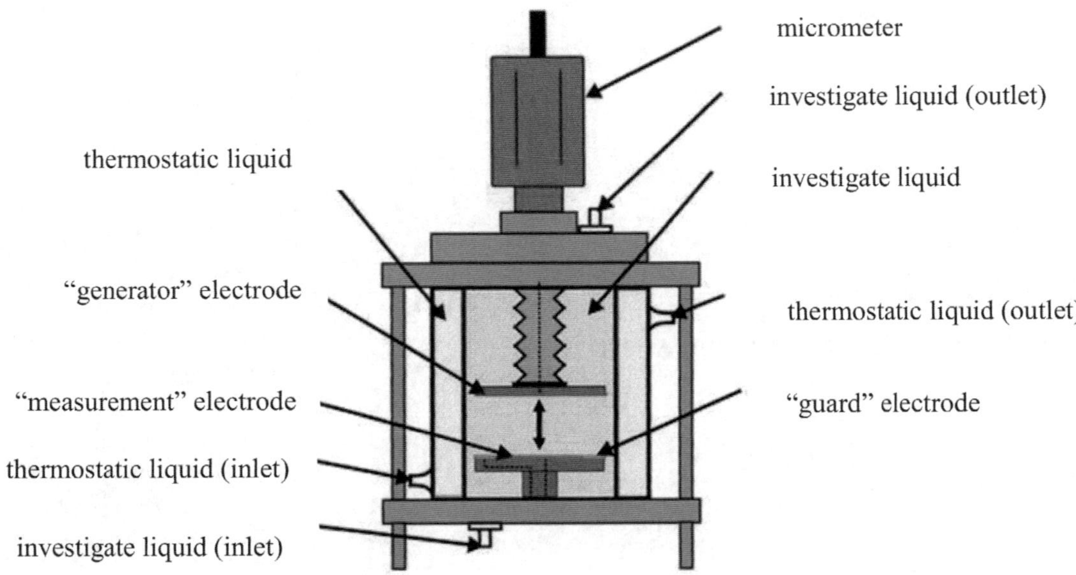

Figure 1. Conductivity cell model

Journal of Physics: Conference Series **1420** (2019) 012021 doi:10.1088/1742-6596/1420/1/012021

Development and validation of secondary method for pH measurement of buffer solutions

Lyudmila Dimitrova, Boryana Koleva

Bulgarian Institute of Metrology, Sofia, Bulgaria

Abstract. This paper presents the development of a secondary pH measurement standard and the design of the most critical part of the measurement standard - the secondary cell. The parameters affecting the measurement accuracy are discussed. The validation of the method together with a full uncertainty budget is described. The intended use of the secondary pH measurement standard by the Bulgarian Institute of Metrology (BIM) is also presented.

Since pH cannot be determined in terms of a fundamental (or base) unit of any measurement system, traceability to the International System of Units (SI) can be established using method (called primary method) that operates according to a well-defined measurement equation in which all of the variables can be determined experimentally in terms of SI units. Any limitation in the determination of the experimental variables, or in the theory, must be included within the estimated uncertainty of the method. The primary method for the absolute measurement of pH which is internationally recognized, is described in the IUPAC Recommendations 2002 [1] and is based on Harned cell (cell without transference). This method provides pH values of buffer solutions directly traceable to SI (primary buffer solutions) but in most applications its expensive use is not justified. A secondary method for pH measurement using cell with transference (secondary cell or differential potentiometric cell) is also internationally recognized and provides traceability to SI. The secondary method is used for calibration of lower level (secondary) buffer solutions through primary buffer solutions. The procedure is simpler and faster and requires cheaper equipment.

In this presentation the development of a secondary standard applying the secondary method for pH measurement is described. The design of the most critical part of the measurement standard, the secondary cell, is presented. The examination of the parameters affecting the measurement accuracy which was done using three quasi-identical buffer solutions is discussed. The measurement parameters were controlled, registered and the results processed by in-house developed specialized software. The secondary method was validated using primary buffer solutions produced by the Bulgarian Institute of Metrology (BIM), certified by the primary pH measurement standard. The outcome of the validation process is presented. The standard was used for calibration of secondary buffer solutions and the results and a full uncertainty budget are discussed.

The developed secondary pH measurement standard will be used for production of secondary buffer solutions which will be at lower cost but with similar quality compared to the primary ones. The solutions will be offered to field laboratories for calibration of pH-meters and used by BIM as reference samples for interlaboratory comparisons and proficiency tests.

This work is part of the activities included in the project 16RPT01 ChemMet-Cap funded by the European Metrology Program for Innovation and Research (EMPIR).

References.

[1] R P Buck et all 2002 Measurement of pH. Definition, Standards, and Procedures (IUPAC Recommendations 2002) *Pure Appl. Chem* **74** (11) 2169

Journal of Physics: Conference Series **1420** (2019) 012022 doi:10.1088/1742-6596/1420/1/012022

Development of optical nanosensors based on dynamic light scattering and fluorescence and detection devices for them

A Yu Sadagov, M K Alenichev, E B Drozhzhennikova, A A Yushina, A D Levin and A I Nagaev

All-Russian Scientific Research Institute of Optical and Physical Measurements (VNIIOFI), Ozernaya 46, Moscow, 119361, Russia

e-mail: levin-ad@vniiofi.ru

Abstract. The paper shows the research work at VNIIOFI in the field of optical nanosensors based on dynamic light scattering (DLS) and fluorescence. Both nanosensors and diagnostic devices for them were developed.

Optical nanosensors provide quick and instrument- simple determination of analytes interest for clinical diagnostics, food control and environmental monitoring. Recently, they began to attract special attention in the case of point of care diagnostics. Inexpensive and reliable measuring instruments for off-laboratory analyzes are also required in different fields, for example, for food control. A significant part of optical nanosensors is used to determine the concentration of analytes directly in liquids. In case of single-phase analysis, probe nanoparticles are inject into the liquid, the optical response from which changes significantly as a result of interaction with the analyte. The study on optical nanosensors of based on dynamic light scattering (DLS) [1] and fluorescence [2] is carried out at VNIIOFI. Both nanosensors and diagnostic devices for them were developed.

In DLS nanosensors, nanoparticles functionalized with receptor molecules capable of selectively binding to analyte molecules are used as probes introduced into the liquid. Using DLS, the change in the hydrodynamic diameter D_H of such particles as a result of aggregation due to interaction with the analyte is measured. In the conventional DLS sensors, the addition of analyte molecules promotes aggregation; therefore, an increase in D_H is proportional to the concentration of molecules. This approach is especially effective for large-size analytes (for example, viruses or macromolecules). We have implemented an alternative approach, in which part of the nanoparticles is functionalized by receptor molecules, and the other part by the analyte molecules themselves. At the first stage of the analysis, nanoparticles functionalized by receptors are mixed with the solution being analyzed, and for a short time, some of the receptors are bound by analyte molecules. In the second stage, nanoparticles functionalized with analyte molecules are added. With this approach, the presence of analyte molecules in the analyzed solution inhibits aggregation and the hydrodynamic diameter decreases with increasing analyte concentration. It is shown that this approach allows us to analyze small molecules and makes it possible to significantly expand the range of determined concentrations. A nanosensor for determining chloramphenicol (an antibiotic, which is strict controlled in food products), developed according this idea, is considered.

A fluorescence-based nanosensor has also been developed to detect restored glutathione, a non-protein low molecular weight compound whose concentration in the blood plasma is the object of interest in the diagnosis of a number of diseases. Colloidal quantum dots (QD) CdSe / ZnS were used as fluorophores. Methyl viologen was applied to the surface of the QD (through a substance — linker), whose molecules "turn off" the QD fluorescence due to nonradiative transfer of the excitation energy. If there are glutathione molecules in the solution, they replace the linker substance; as a result, the methyl viologen – QD bond weakens and the "switched off" fluorescence is restored in proportion to the concentration of glutathione.

Based on the principles discussed above, nanosensors for other analytes can also be developed. For their effective use, fairly simple, reliable and small-sized devices that measure DLS or fluorescence are needed. Such measuring instruments have also been developed.

A DLS-based detection device is a small-sized DLS analyzer that allows, in addition to the D_H, to measure the number concentration of nanoparticles, as well as the dimensional parameters of nonspherical nanoparticles.

As a light source, a self-made laser based on a laser diode is used, the wavelength is 640 nm, it is possible to adjust the laser output power. The light transmitted through the sample is incident on the photodetector of transmitted light, which provides a measure of absorbance which is necessary to determine the number concentration of nanoparticles. The light scattered by the sample at an angle of 90 ° passes through a film dichroic polarizer, which is installed in a rotary device and is necessary for measuring the dimensional parameters of nonspherical nanoparticles by the method of multipolarization light scattering. After that, the light is collected by a lens, which with a decrease transmits the image to the optical fiber. Before reaching the optical fiber, the light passes through the aperture, which limits the scattered light collection area and increase the coherence factor. Further along the optical fiber, the light enters the photon counting module, which acts as a scattered light detector. The software correlator is used for the calculation of the scattered light intensity.

A fluorescence-based detection device uses the concept of a filter fluorimeter.

It provides for the use of laser diodes as sources of quasi-monochromatic light for excitation of fluorescence, measurement of emission intensity in narrow spectral ranges selected by interference filters. The use of laser diodes makes it possible to provide a sufficiently high power of exciting light, which makes it possible to use a photodiode instead of a PMT as a photodetector.

The light from the source with an aspherical plastic lens is focused in the center of a standard fluorometric cell located in a thermostatically controlled cell compartment. There are two channels for collecting emission light at an angle of 90° to the direction of the exciting light; in each of these channels, light is collected by plane-convex spherical lenses and interference filters. At the same time, it is possible to measure the emission intensity simultaneously at two wavelengths, this is necessary for detecting a signal from sensors using the ratio of emission intensities at two different wavelengths (the so-called ratiometric) as an analytical signal. An example are sensors based on fluorescence resonance energy transfer (FRET).

The device designed for glutathione sensors used a laser diode with a power of about 100 mW, the with the emission spectrum in the range from 456 to 458 nm, and interference filters at a wavelength of 610 nm (maximum emission of CdSe / ZnS quantum dots). The design of the detecting device provides for the possibility of a simple replacement of the light source and interference filters if it is necessary to detect signals from nanosensors with other analytes.

The work was supported by the Ministry of Science and Higher Education of Russian Federation within the framework of the implementation of agreement No. 14.624.21.0045 of September 26, 2017 (unique identifier RFMEFI62417X0045).

References.

[1] A D Levin, M P Nikitin, M K Alenichev et al. 2019 Nano-biosensors based on dynamic light scattering Proc. SPIE 11060, Optical Methods for Inspection, Characterization, and Imaging of Biomaterials IV, 110600Y Doi: 10.1117/12.2526055

Journal of Physics: Conference Series **1420** (2019) 012022 doi:10.1088/1742-6596/1420/1/012022

[2] A. D. Levin, A Yu Sadagov, M. K. Alenichev et al. 2019 Fluorescent nanosensors on colloidal quantum dots *V International Conference "Laser, Plasma Research and Technology". Collection of scientific papers* (Moscow: NRNU MEPhI) **1** 348

Journal of Physics: Conference Series **1420** (2019) 012023 doi:10.1088/1742-6596/1420/1/012023

Perfecting technology for producing thermistors at VNIIFTRI

Priadka A.A

All-Russian Scientific Research Institute of Physicotechnical and Radio Engineering Measurements (VNIIFTRI), Mendeleevo, Moscow region, Russia, 141570

e-mail: priadka@vniiftri.ru

Abstract. In this work characteristics of the thermistors, created by the method developed in VNIIFTRI were investigated. Parameters of thermistors are studied, dependences of their properties on manufacturing techniques are revealed.

Task this work is debugging of the production technology and also studying properties of thermistor depending on parameters of their manufacture. It will allow to improve characteristics of wattmeter on their basis, processibility production and also to reduce marriage percent.

The objective of this work is to debug production technology, as well as studying the properties of temperature resistors, depending on the parameters of their manufacture. This will improve the performance of power meters based on them, the manufacturability production, as well as reduce the percentage rejects.

Measuring the power electromagnetic waves is one of the main types of measurements in the microwave range. Measurement microwave power using thermistor converters is carried out by replacing the power of high-frequency electromagnetic waves absorbed in a thermistor by direct current power. This ensures the traceability of the microwave power unit to the standards of the basic physical quantities - voltage and DC resistance. Also, a feature thermistor converters is high linearity and stability over time [1].

For this reason, thermistor converters underlie almost all national microwave power standards of leading foreign countries [2].

In the Russian Federation, to date, high-frequency thermistors have not been produced. In the USSR, several types of thermistors were produced for use in power converters, but the upper frequency of commercially available coaxial thermistor converters did not exceed 12 GHz. The production technology of high-frequency thermistors in the Russian Federation has been lost. This circumstance makes it difficult to improve the existing state primary, secondary and working microwave power standards, and hinders the development of new generation working microwave power standards of the through-type (power calibrators). This leads to high relevance of the work.

The properties of a thermistor are affected by many parameters: the content of each of the three components its composition, the type of plasticizer burned, the heating rate the furnace, the annealing temperature, time, aging temperature, and many other parameters. Changing only one of them changes the properties of products.

The manufacturing of the first batches of thermistors was carried out in a short time and by limited forces, and the production cycle of one batch of products takes at least a week. Due to these reasons, it

Journal of Physics: Conference Series **1420** (2019) 012023 doi:10.1088/1742-6596/1420/1/012023

was not possible to investigate the influence of all technological parameters, and they were not optimized.

In this paper, we studied the influence the following technological parameters:

- manganese content in the working fluid;
- maximum annealing temperature;
- furnace heating rate;
- aging time.

The development of three-component systems is complicated by the fact that much more experiments are required to study their properties. Therefore, due to the complexity and high cost the production process at this stage, it was decided limit ourselves to studying the effect of manganese in the manganese-copper-cobalt system. In the experiment, equal mass parts of copper and cobalt were used, and only the mass concentration of manganese was changed. The result of the experiment can be seen in Figure 1. With a decrease in the mass fraction of manganese, the linear resistance initially decreases, then stabilizes at a certain value.

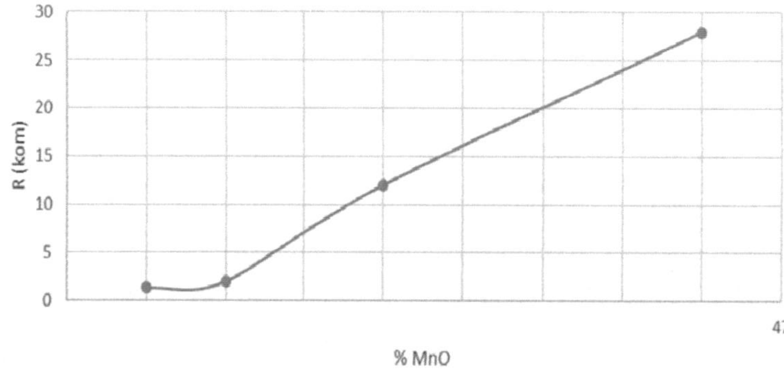

Figure 1. Dependence of the linear resistance of the thermistor on the mass fraction of manganese in its composition

The annealing temperature of thermistors dramatically affects their characteristics and appearance. The composition of the working fluid of a thermistor can include 3 phase components - an unreacted mixture of oxides, an oxide melt and the so-called spinel solid solutions - three-component polycrystalline structures that are semiconductors [3]. Spinel formation begins at a temperature unique to each composition. Therefore, when the percentage composition of the components changed, it was necessary to select the sintering temperature again. When the filament overheats by 10 degrees, it melts, and at a temperature below the optimum by 10 degrees, the working fluid still consists of a sintered mixture of oxides, and is practically an insulator. The furnace heating rate does not play a significant role.

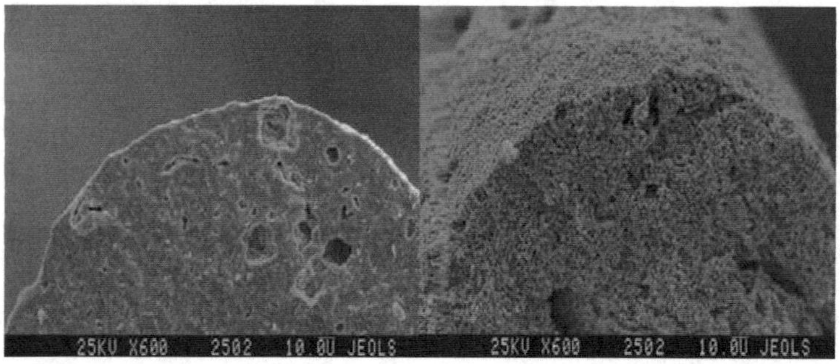

Figure 2. External differences of the working fluid filament when changing the annealing temperature by 30 degrees.

Journal of Physics: Conference Series **1420** (2019) 012023 doi:10.1088/1742-6596/1420/1/012023

It is also planned to study the dependence of the properties of thermistors on the content of cobalt and copper, to study the possibility of accelerating the aging process. To optimize some other technological processes.

As a result of the work, the influence of the listed parameters on the resistance and microstructure of thermistors was established. Charts are constructed that allow the method of interpolation to optimize technological parameters for future products. Refined data obtained in early experiments. The production flowchart has been simplified.

References.

[1] Petukhova N A, Gatchin Yu A 2016 Measurement of power in the microwave range *Successes in modern science and education* **2** (11) 146-151 (In Rus.)

[2] Perepelkin V A, Pivak A V, Chuiko V G 2003 Investigation of the linearity reference thermistor wattmeters and estimation of the minimum achievable measurement errors of the microwave power ratio *Measureament techniques* **10** 45-51 (In Rus.)

[3] Zinovik E V 2008 Thermodynamics and properties of solid solutions of the Fe_3O_4 – Mn_3O_4 – Co_3O_4 system (In Rus.)

Nanoparticle concentration in engine emissions

D I Belenkii, T M Magomedov, D M Balakhanov

All-Russian Scientific Research Institute of Physical-Technical and Radiotechnical Measurements

Abstract. Nanoparticles arising during the operation of the internal combustion engine are dangerous for human health, as they easily penetrate the body through the respiratory tract. Nanoaerosol in car exhaust gases is a complex mixture of different components, often having a polydisperse size distribution of particles, most of which are from 5 to 100 nm. In Europe, since 2008, the Regulation establishes limits of the mass concentration and the counting concentration of nanoparticles in engine emissions (European Commission, Commission Regulation No. 692/2008), however, there is no such Regulation in Russian Federation, which in turn leads to the deterioration of the environmental situation and people health.

1. Introduction

In Russia, all existing regulatory documentation in the field of control of particulate matter in atmospheric air is based on mass concentration measurements. However, in the case of automobile exhausts, most of the particles are in the nanoscale range.

The main problems of the existing approach are related to:

— Rationing of maximum permissible concentrations of solid particles in atmospheric air by measuring their mass concentration leads to the fact that at very small mass concentrations, the number concentration of particles can be extremely high, since more than 90 % of the solid particles in the exhaust are in the nanometer size range.

— Particulate matter is the greatest threat to humans. In the process of breathing, they freely penetrate into the lungs, subsequently causing bronchitis and cancer, and especially small particles act as condensation nuclei and can penetrate into the blood.

— At the moment, particle number concentration in automobile exhausts is not controlled.

2. Nanoaerosol in car exhaust gases

Car exhausts in most cases contain solid particles based on carbon materials, ash and volatile components. The schematic ratio of the components in the exhaust is shown in figure 1. When enriching the combustible mixture, solid carbon particles are formed [1]. A small fraction of the oil and fuel are not oxidized and remain as volatile compounds. The main difference between the exhaust gases of diesel and gasoline engines is that in the case of the latter, the proportion of metallic inorganic particles is much higher due to the spark when igniting the mixture.

One of the most important points is that when sampling and measuring the size and concentration of particles in the exhaust gases, the cooling of the sample is an integral part of the sampling. During cooling, there is transition of volatile materials from the gas phase, which is accompanied by the processes of nucleation, condensation and adsorption on solid particles. This process leads to distortion of the real particle size distribution.

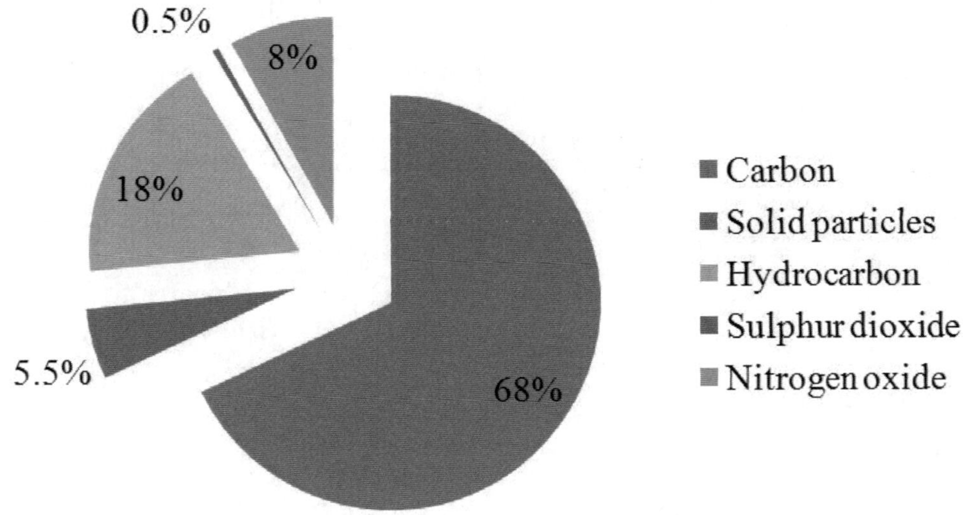

Figure 1. Schematic ratio of the components in the exhaust

Typical engine exhaust size distribution both mass and number weightings are shown in figure 2 [2].

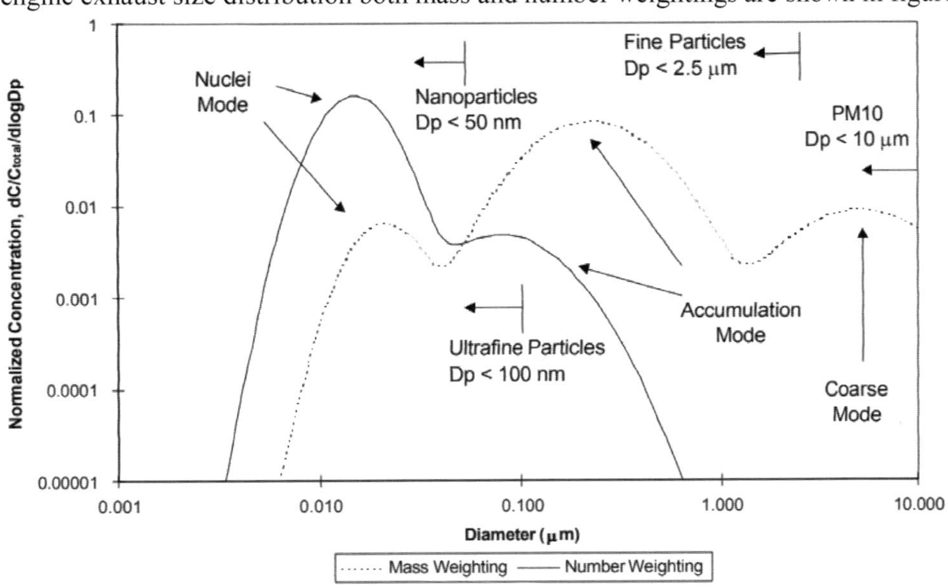

Figure 2. Typical engine exhaust size distribution both mass and number weightings

The largest share of particles in total number concentration are particles with sizes less than 100 nm, while the largest share in the total mass is made by particles of large sizes. When measuring only the mass concentration, the share of nanoparticles is insignificant, and their harm may not be reflected because of the extremely small masses (table 1). However, huge number concentrations and good penetrating ability make them extremely dangerous for human health.

Table 1. Dependence of particles number concentration of the same density and optical properties in 1mcg / m3 on their diameter

Mass concentration, μg/m^3	Particle diameter	Number concentration, m^{-3}
1	1 nm	1,91E+15
1	10 nm	1,91E+12
1	100 nm	1,91E+09
1	1 μm	1,91E+06
1	10 μm	1,91E+03

3. Measurements of nanoparticle size distribution and number concentration

The measurement method is based on the size separation of aerosol particles as they pass through an electric field, where charged aerosol particles change their trajectory depending on their size, aerosol flow rate, electric field strength and the geometry of the classifier [3].

The relationship between differential electrical mobility and particle size for spherical particles is described by the equation (1):

$$Z = \frac{NeS_C}{3\pi\mu d} \qquad (1)$$

Where:
N = number of elementary charges per particle;
e =elementary charge;
μ = gas dynamic viscosity;
Sc = slip correction factor;
d = particle diameter.

The slip correction is based on the frictional force (Stokes ' law) acting on a spherical particle of very small size moving with a low Reynolds number in a gas medium. The correction is approximated by the equation (2):

$$S_C = 1 + K_n\left[A + Bexp\left(-\frac{C}{K_n}\right)\right] \qquad (2)$$

Where:
A, B, C = empirical constants;
K_n = Knudsen number.

The dependence between the differential electric mobility of the particle and the parameters of the cylindrical classifier is described by the equation (3):

$$Z = \frac{q_{sh}}{2\pi UL}\ln(\frac{r_2}{r_1}) \tag{3}$$

Where:
q_{sh} = aerosol flow rate in classifier;
r_1 = outer radius of the inner cylinder of classifier;
r_2 = inner radius of the outer cylinder of classifier;
U = DC voltage creating an electric field;
L = effective length between inlet and outlet of aerosol stream in classifier.

According to the described equations can be obtained (4):

$$\frac{d}{N_e} = \frac{2S_cUL}{3\mu q_{sh}\ln(\frac{r_2}{r_1})} \tag{4}$$

This equation allows to define the size of the particle leaving the electrostatic field of the classifier if the number of elementary charges on the particle is known.

4. Regulations for particle number concentration and size measurements in automobile exhaust

Based on the above, it becomes clear that it is imperative to control particulate matter in automobile exhausts. To date, the Russian Federation has A Technical Regulation of the Customs Union "On the safety of wheeled vehicles" TR CU 018/2011, in accordance with which only 2 emission parameters are regulated: level of CO and CH and opacity [4]. These parameters can not fully characterize the harmfulness of automobile exhausts, since they absolutely do not take into account number concentration and size of solid particles.

Commission Regulation No. 6692/2008 was applied in the European Union in 2008 [5], according to which, in addition to gas levels, mass and number concentration of particles in automobile exhausts is monitored (table 2).

Table 2. Euro 5 Emission Limits

			Limit values													
		Reference mass (RM) (kg)	Mass of carbon monoxide (CO)		Mass of total hydrocarbons (THC)		Mass of nonmethane hydrocarbons (NMHC)		Mass of oxides of nitrogen (NO_x)		Combined mass of hydrocarbons and oxides of nitrogen (THC + NO_x)		Mass of particulate matter (1) (PM)		Number of particles (2) (P)	
			L_1 (mg/km)		L_2 (mg/km)		L_3 (mg/km)		L_4 (mg/km)		$L_2 + L_4$ (mg/km)		L_5 (mg/km)		L_6 (#/km)	
Cat.	Class		PI	CI	PI	CI	PI	CI	PI	CI	PI	CI	PI	CI	PI	CI
M	-	All	1000	500	100	-	68	-	60	180	-	230	5.0/4.5	5.0/4.5	-	$6.0 \cdot 10^{11}$
	I	RM ≤ 1 305	1000	500	100	-	68	-	60	180	-	230	5.0/4.5	5.0/4.5	-	$6.0 \cdot 10^{11}$
N_1	II	1 305 < RM ≤ 1 760	1810	630	130	-	90	-	75	235	-	295	5.0/4.5	5.0/4.5	-	$6.0 \cdot 10^{11}$
	III	1 760 < RM	2270	740	160	-	108	-	82	280	-	350	5.0/4.5	5.0/4.5	-	$6.0 \cdot 10^{11}$
N_2	-	All	2270	740	160	-	108	-	82	280	-	350	5.0/4.5	5.0/4.5	-	$6.0 \cdot 10^{11}$

Key: PI = Positive Ignition, CI = Compression Ignition
([1]) A revised measurement procedure shall be introduced before the application of the 4,5 mg/km limit value.
([2]) A new measurement procedure shall be introduced before the application of the limit value.
([3]) Positive ignition particulate mass standards shall apply only to vehicles with direct injection engines.

Taking into account that in the territory of the Russian Federation there is State Primary Standard for the units of disperse parameters of aerosols, suspensions, and powder materials GET 163 which provides metrological assurance for parameters of solid particles in automobile exhausts, it will be appropriate to update the list of parameters regulated in accordance with TR CU 018/2011.

5. References.

[1] Kittelson D B 1998 Engines and nanoparticles: a review. *J. Aerosol Sci.* **29** (5/6) 575-588
[2] Kittelson D B 2000 Nanoparticle Emissions from Internal Combustion Engines. *The Royal Society Discussion Meeting Ultra Fine Particles in the Atmosphere*
[3] ГОСТ Р 8.775-2011 "ГСИ. Дисперсный состав газовых сред. Определение размеров наночастиц по методу дифференциальной электрической подвижности аэрозольных частиц"
[4] ТР ТС 018/2011 Технический регламент Таможенного союза "О безопасности колесных транспортных средств"
[5] Commission Regulation (EC) No 692/2008. Official Journal of the European Union

Journal of Physics: Conference Series **1420** (2019) 012025 doi:10.1088/1742-6596/1420/1/012025

Detection of synthetic dyes in red wines

A Aprelev, E Davydova , V Smirnov, A Pryadka , E Lavrov

All-Russian Scientific Research Institute of Physicotechnical and Radio Engineering Measurements (VNIIFTRI), Mendeleevo, Moscow region, Russia, 141570

Abstract. The falsification of drinks, syrups, extracts and concentrates, as a rule, is of a technological nature and is carried out by diluting, as well as completely or partially replacing natural ingredients with essences, sweeteners, colorants and other additives, including synthetic ones. Painted fruits are distinguished by the presence of anthocyanins - compounds that determine their color and have high antioxidant activity. The paper describe developments of libraries of spectra of artificial dyes used as additives in drinks and food products and of spectra of natural juices (containing anthocyanins) in the wavelength range of the visible part of the spectrum from 400 to 760 nm.

Currently, an urgent problem is the identification of falsifications of wine and fruit juices entering the consumer market. Many studies have found that the main source of free radicals and antioxidants in the human body, due to the high content of phenolic compounds, are alcoholic beverages (wines), tea, coffee, juices, vegetables and fruits. The insufficiency of food control methods does not allow for timely detection of deviations from the norm and the detection of counterfeit products. The strategy to improve the quality of food products in the Russian Federation until 2030 and current market trends require the establishment of additional criteria for the quality of products, including their identification features, which should be preceded by serious scientific work. Only after this is it possible to develop measurement methods and techniques with the participation of specialists from different fields [1].

Painted fruits are distinguished by the presence of anthocyanins - compounds that determine their color and have high antioxidant activity.

Anthocyanins are colored plant glycosides containing, as an aglycon, anthocyanides-substituted 2-phenylchromenes related to flavanoids. Anthocyanins have several forms of isomers. An experimental study was conducted at a pH of 3 to 4, which corresponds to the pH of natural red wine [3].

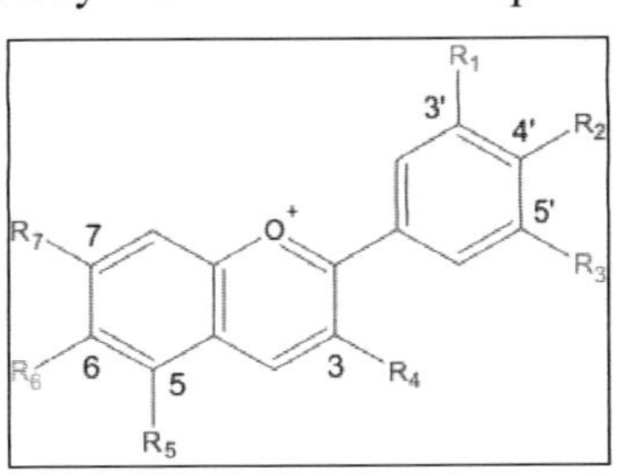

Figure1. Total structure of Anthocyanidins.

The color of anthocyanins and anthocyanidins is determined by their sensitivity to pH: in an acidic environment (pH <3), anthocyanins (and anthocyanidins) exist in the form of red pyrilide salts; when the pH is increased to 4-5, the hydroxide ion joins to form a colorless pseudo base; with a further increase in pH to ~ 6-7, water is split off to form a blue quinoid form, which, in turn, splits off a proton at pH ~ 7-8 to form a purple phenolate, and finally, at pH above 8, the quinoid phenolate hydrolyzes with a rupture of the chrome cycle and the formation of the corresponding yellow chalcone.

The falsification of drinks, syrups, extracts and concentrates, as a rule, is of a technological nature and is carried out by diluting, as well as completely or partially replacing natural ingredients with essences, sweeteners, colorants and other additives, including synthetic ones.

Table 1. General structure of anthocyanidins

Anthocyanin	R_1	R_2	R_3	R_4	R_5	R_6	R_7
Aurantinidin	-H	-OH	-H	-OH	-OH	-OH	-OH
Cyanidin	-OH	-OH	-H	-OH	-OH	-H	-OH
Dolphinidin	-OH	-OH	-OH	-OH	-OH	-H	-OH
Europinidine	$-OCH_3$	-OH	-OH	-OH	$-OCH_3$	-H	-OH
Luteolinidine	-OH	-OH	-H	-H	-OH	-H	-OH
Pelargonidine	-H	-OH	-H	-OH	-OH	-H	-OH
Malvidin	$-OCH_3$	-OH	$-OCH_3$	-OH	-OH	-H	-OH
Peonidine	$-OCH_3$	-OH	-H	-OH	-OH	-H	-OH
Petunidine	-OH	-OH	$-OCH_3$	-OH	-OH	-H	-OH
Rosinidine	$-OCH_3$	-OH	-H	-OH	-OH	-H	$-OCH_3$

To detect these falsification methods, the mass concentration of the reduced extract, i.e. the sum of all extractive substances minus sugar. [2]. The main physical and chemical indicators used during the examination include the mass fraction of solids (%), acidity and resistance.With qualimetric identification and quality examination, the mass fraction of solids and acidity are determined, the reliability of information on the composition of the raw material is confirmed by examining the presence of dyes, preservatives, and sweeteners. The use of natural ingredients in the composition of the raw materials is confirmed on the basis of studies of characteristic groups of compounds (organic acids, sugars, anthocyanins, flavonoids, etc.) inherent in a particular type of natural raw material.

Studies conducted in recent years in our country and abroad allow us to conclude that the anthocyanin profile can be considered "fingerprints" of pigment-stained fruits. According to studies, the main anthocyanin of the staining fruit is cyanidin-3-O-glucoside.

VNIIFTRI together with the Federal State Budget Scientific Institution "V. Gorbatov Federal Scientific Center for Food Systems RAS has developed a method for determining the presence of anthocyanins in biotechnological environments (Registration number of patent application 2018145610).

VNIIFTRI also developed the Detection of the presence of synthetic dyes in red wines by optical density. Advantages of the proposed method:

- rapid analysis, which allows one minute to detect the synthetic dye in the drink;
- high profitability and, accordingly, a small contribution of the measurements made to the unit cost of the controlled product (economic feasibility);
- reproducibility at any stage of production, bottling, transportation and sale of the drink.

An analytical result is achieved by spectrophotometry of the test sample and comparative analysis of the characteristic curves obtained during spectrophotometry, a natural sample of cyanidin-3-O-glucoside, as well as a number of synthetic dyes, which increases the accuracy of the analysis taking into account the specific conditions of the test.

A group of specialists of VNIIFTRI conducted a study of changes in the extrema of the optical density of natural wine at pH values from 1 to 10. The resulting wave graphs of the function of the optical density of natural wine samples at various pH values propagating along the Ox axis, plotted at a fixed point in time (t = const) have all types of shift of the absorption band (Figure 1).

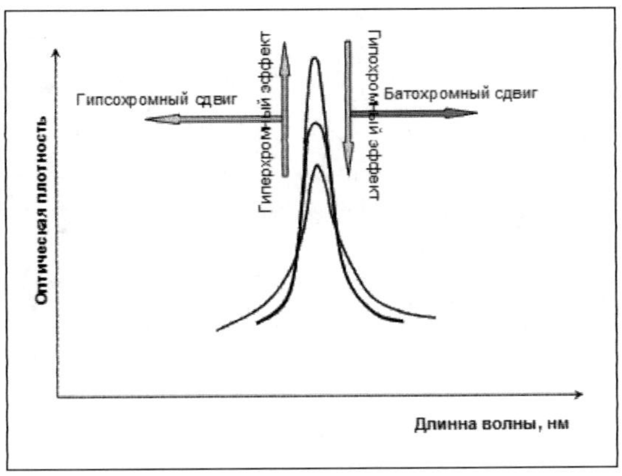

Figure 2. Types of absorption band shifts.

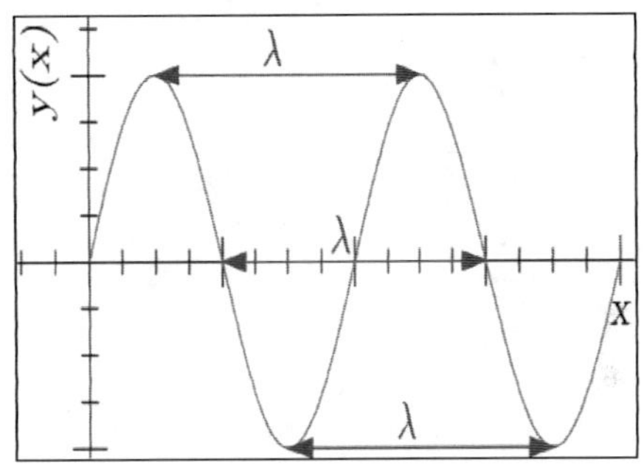

Figure 3. Methods of measuring the wavelength λ.

Hyperchromic and hypochromic shifts - increase and decrease in intensity, respectively. An increase in band intensity is called the hyperchromic effect, and a decrease is called the hypochromic effect.

With increasing solvent polarity due to an increase in the pH of a natural wine solution, the long-wavelength absorption band undergoes a bathochromic shift. The magnitude of the bathochromic and hypsochromic shift was determined by measuring as the distance between a pair of neighboring maxima y (x). All data for natural wine are shown in Table 1, for synthetic dyes, all values of the absorption band offsets are shown in Table 2.

Table 2. Table of Optical density depending on the pH value of the investigated sample of natural wine.

Sample number	pH	Extremum, nm	Bathochromic shift, nm	Hypsochrome shift, nm	Hyperchromic shift, nm	Hypochromic shift, nm
1	1	410	-	13,0	Extraction coefficient of investigated wine varies from 45 to 52	
2	2	414	-	9,0		
3	3	420	-	3,0		
4	3,5	423	-	-		
5	4	704	281,0	-		
6	5	785	362,0	-		
7	6	720	307,0	-		

According to normative documents concerning the quality of natural red wines (GOSTs and TU), the value of the hydrogen index of natural wine ranges from 3.0 to 4.0. Graphs of the dependence of

Journal of Physics: Conference Series **1420** (2019) 012025 doi:10.1088/1742-6596/1420/1/012025

optical density on the wavelength, as well as the magnitude of the shift of the absorption band for all types of shifts are considered as criteria for determining the naturalness of wine.

Table 3. Table of Optical density depending on the pH value of the studied samples of synthetic dyes.

Sample number	pH	The value of the extremum, nm	Bathochromic shift, nm	Hypsochrome shift, nm	Hyperchromic shift, nm	Hypochromic shift, nm
11	3,5	423	-	-	-	-
12	3,5	431	-	9	12	-
13	3,5	410	12	-	-	8
14	3,5	380	-	43	75	-
15	3,5	411	-	12	-	22
16	3,5	515	88	-	-	37
17	3,5	413		11	-	25
18	3,5	472		49	-	25
19	3,5	405	18		-	38

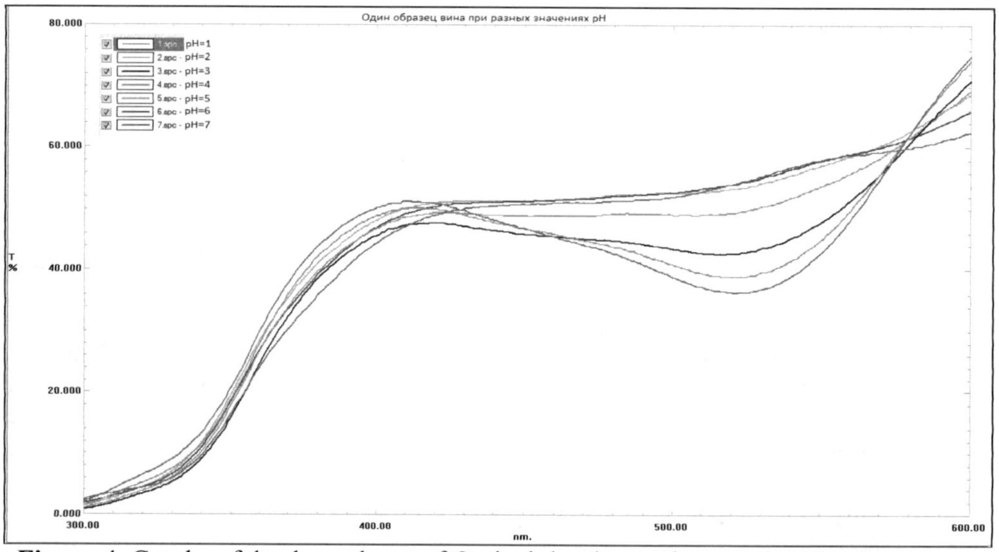

Figure 4. Graphs of the dependence of Optical density on the wavelength of a sample of red wine at various pH values.

On Figure 4 plots of the dependence of optical density on the wavelength of natural wines of Cabernet and Isabella and dyes E 122, E 180 and E 124. Dye plots are outside the specified range.

As show the Figure 5 the spectrogram of the synthetic dye does not change at different pH values.

Red wines from Isabella grapes contain diglucoside malvidol, which causes a hypsochromic shift and the extremum of the optical density is at 411 nm, while the extremum of Cabernet wine is at 423 nm ± 2, that is, it falls into the specified wavelength range. The peculiarity of the chemical composition of wine from Isabella grapes must be known and taken into account when analyzing wine samples for naturalness.

In the course of the work, natural red wine was falsified in the laboratory by adding natural dyes (chokeberry and natural dye E 163, obtained from Isabella grape skins). In the study of falsifications, spectra similar in appearance were obtained, but the extrema are different. Dye E163 does not have an extremum in the characteristic wavelength region of 410-425 nm, while natural wine has an extremum at 423 nm. (Figure 5).

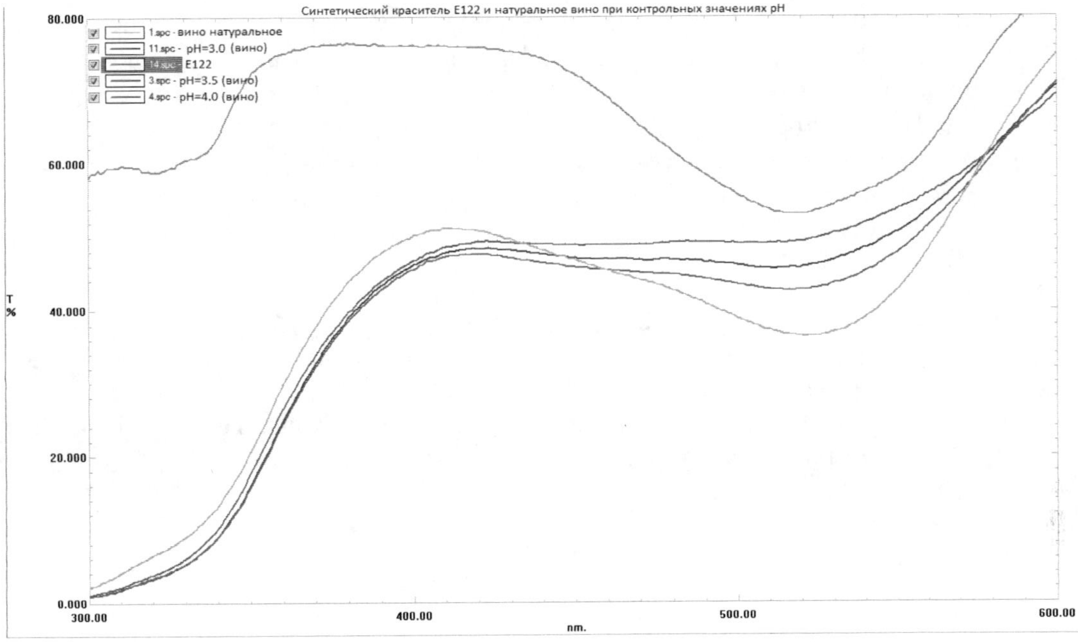

Figure 5. Synthetic dye E122 and natural wine at control pH values.

Extreme graphs of the optical density of red wines with synthetic dyes at pH 3.0; 3.5 and 4.0 are outside the range of 410-425 nm.

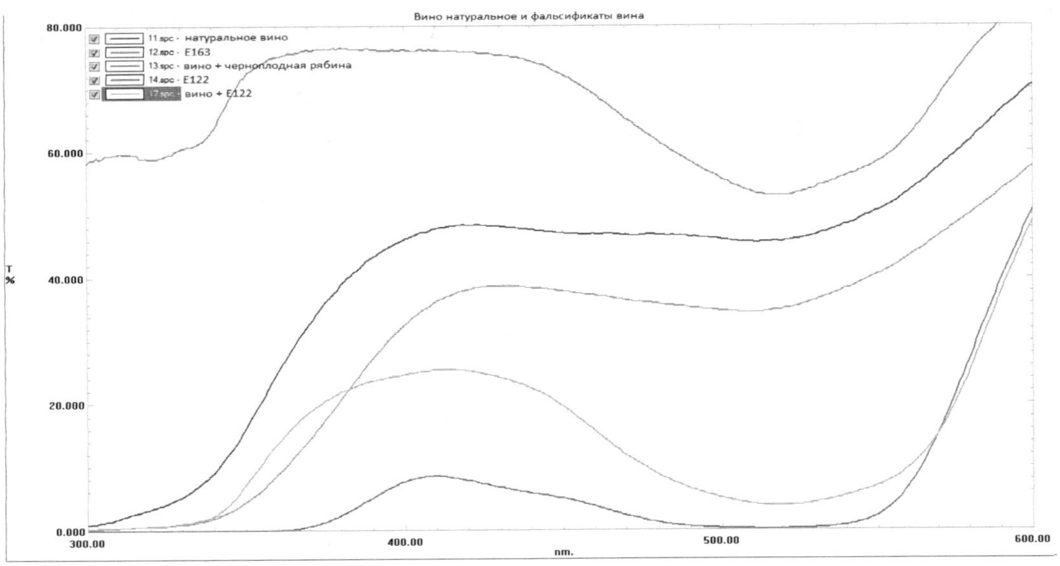

Figure 6. Natural wine and wine falsifications.

Journal of Physics: Conference Series **1420** (2019) 012025 doi:10.1088/1742-6596/1420/1/012025

Thus, it has been established that natural red wines, corresponding in terms of quality to the normative documents for wine, and in particular in terms of pH, have characteristic spectra and an extremum in the wavelength range from 410-425 nm. All extremum values of the optical density of the test sample that are not in this wavelength range (for the bathyromic and hypsochrome shifts) show the presence of a synthetic dye in the test wine sample. For pink and white wines, similar studies are being conducted, the results of which will be reported later. The spectrograms for natural juices containing anthocyanins, which provide the juice with a red color and juices containing dyes, were similarly studied. The results are shown in Figure 6. For natural juices, the extremum of the optical density is also in the wavelength range of 410-425 nm.

Conclusions.

- A library of spectra of artificial dyes used as additives in drinks and food products have been created; and a library of spectra of natural juices (containing anthocyanins) in the wavelength range of the visible part of the spectrum from 400 to 760 nm have been created as well
- A fast, low-labor and cost-effective method for assessing the naturalness of drinks based on spectrophotometry in comparison with HPLC, mass spectrometry and other expensive analysis methods has been developed.
- A patent application has been filed.
- A measurement technique (MVI) is being developed

References.

[1] Hovhannisyants L A, Khurshutdyan S A, Galstyan A G 2018 Monitoring the quality of food products is a basic element of the strategy, *Quality Control of Products*, **4**

[2] Panasyuk A L, Slavskaya S L, Egorova O S, Kharlamova L N 2006 Antioxidant properties and physiological value of wines from red and aronia, *Storage and processing of agricultural raw materials*, **12**, p. 41-44

[3] Chupakhin G N, Maslennikov P N, Skrypnik P N 2011 Natural antioxidants. (Environmental aspect), Monograph, Publishing House of the Baltic Federal University named after Immanual Kant

[4] Tsipko T G 2012 Analytical decisions in determining some indicators of safety and quality of food products, Abstract of dissertation for the degree of Doctor of Chemistry (Krasnodar, Russia)

[5] Aline Augusti Boligon 2014 Technical Evaluation of Antioxidant Activity, *Medical chemistry* 517-522

Research of metrological characteristics of reference installations based on helium picnometry and hydrostatic weighing methods

Egor P Sobina

Ural Research Institute for Metrology (UNIIM), Ekaterinburg, Russia

e-mail: 251@uniim.ru

Abstract. The results of investigations of metrological characteristics of reference installations based on the helium pycnometry and hydrostatic weighing methods are demonstrated.

The creation of new porous materials, the quality of which should be controlled, leads to the emergence of a wide stock of instruments for measuring porosity characteristics. Important parameter of rocks is their open porosity and gas permeability. Porosity is a property of rocks, determined by the presence of voids in them: pores, cracks and cavities containing oil, gas and water. Thus, the rock open porosity determines the ability of the rock to contain fluids (oil, gas, water) and, therefore, this parameter evaluates both the total hydrocarbon reserves and the oil and gas production technology and well flow rates and productivity. The Order of the Ministry of Natural Resources and Environment of the Russian Federation of 15.05.2014 No. 218 provides for monitoring the rock porosity in accredited laboratories. In this regard, open porosity is studied by almost all core research laboratories, and its measurements are included in the list of mandatory. Currently, open porosity measurements are made using instruments in which measurements are performed using inert gases (most often nitrogen or helium) based on the Boyle-Marriott law. For such measurement instruments (MI), maximum accuracy in determining the metrological characteristics of the MI is obtained by using of reference materials (RMs) for porosity of solids and materials (simulators), which are hollow cylinders made of stainless steel. Also, other measurement instruments are used for measuring open porosity on the basis of various methods: liquid saturation, NMR, electromagnetic, sorption, hydraulic, etc. It is impossible to use RMs, based on the hollow stainless steel cylinders for these methods. That is why, for metrological assurance in the field of open porosity measurements of rocks, rock RMs (simulators) based on stainless steel and alumina cylinders were developed. Research was carried out during the work on the improvement of GET 210-2014 the State Primary Measurement Standard for specific adsorption units of gases, specific surface, specific volume and pore size of solids and materials [1].

The method helium picnometry is based on the change in gas pressure in the measuring chamber of the pycnometer as a result of the expansion of this gas from the auxiliary chamber to the measuring chamber with the sample under study. In the course of the experiment, the pressure sensor readings are taken before and after connecting the chambers, from which the the solid phase volume $V_{\text{solid phase}}$ of the sample under study is calculated. With the help of a micrometer, the length and diameter of the sample are measured, and its volume V_0 is found. By subtracting the volume of its solid phase from the sample volume V_0, the pore volume V_{pore} is found. The open porosity A, % is calculated by finding the ratio of pore volume V_{pore} to sample volume V_0. The hydrostatic weighing method was used only for stainless steel cylinders. To calculate the open porosity, the cylinder volume was determined before and after drilling by hydrostatic weighing method based on Archimedes law with amendments to the buoyancy of air. The open porosity for both described methods was estimated by the equation:

$$A = \left(\frac{V_{пор}}{V_0}\right) \cdot 100 = \left(\frac{V_0 - V_{тв.ф.}}{V_0}\right) \cdot 100 \qquad (1)$$

where

V_0 is the sample volume, cm^3;

$V_{solid\ phase.}$ is the volume of the solid phase, cm^3;

V_{pore} is the pore volume, cm^3.

As a result of this work, two RM sets for open porosity in a wide variation range of porosity (3-50)% were developed, which allow providing metrological support for measurement instruments operating on different principles of operation. The developed RM sets for open porosity are registered in the Federal Information Fund under the following numbers GSO 11116-2018 ÷ GSO 11119-2018 and GSO 10583-2015 [2].

Table 1. Metrological characteristics of RMs for open porosity of rocks (simulators) (OPGP RM UNIIM set) GSO 11116-2018 ÷ GSO 11119-2018

Characteristic \ RM index	OPGP-5	OPGP-15	OPGP -30	OPGP-45
Certified value of open porosity A, %	4,81	16,53	29,64	43,36
Standard uncertainty of Type A u_A, %	0,07	0,17	0,11	0,10
Standard uncertainty of Type B u_B, %	0,04	0,04	0,04	0,04
Combined uncertainty u_C, %	0,08	0,17	0,11	0,11
Expanded uncertainty at k=2 U, %	0,16	0,34	0,23	0,21

Table 2. Metrological characteristics of RMs for open porosity of solids and materials (simulators) (OPTV RM UNIIM set) GSO 10583-2015

RM index	Certified value of open porosity, %	Expanded uncertainty of the certified value (k=2), %	The boundaries of the absolute error of the certified value at P = 0.95,%
OPTV-1	5,15	0,05	± 0,05
OPTV -2	9,99	0,05	± 0,05
OPTV -3	20,30	0,05	±0,05
OPTV -4	30,03	0,05	±0,05
OPTV -5	50,65	0,05	±0,05

The developed RM sets for open porosity GSO 10583-2015 and GSO 11116-2018 ÷ GSO 11119-2018 successfully passed approbation in the process of testing in order to approve 7 types of MI: instruments for measuring porosity "Poromer" (No. FIF 62233-15), instruments for measuring the gas permeability coefficient Darsimeter (No. FIF 59887-15), analyzers of the porosity and gas permeability of rocks MG2P500 (No. FIF 68446-17), analyzers of gas permeability and porosity PIK-PP (No. FIF-73989-19), gas volumetric porosimetes PIK-P, gas permeability and open porosity PPMmeter, relaxometers iTMR-10.

References.

[1] Order of the Federal Agency for Technical Regulation and Metrology dated January 29, 2015 № 128 "On approval of the State primary measurement standard of units of specific adsorption of gases, specific surface, specific volume and pore size of solid substances and materials."

[2] Sobina E P 2016 Development of certified reference material set for open porosity of solid substances and materials (imitators) J. *Reference materials* **2** p 36-43
(In Russ.) Doi:10.20915/2077-1177-2016-0-2-36-43

Singularity of GOST R 57165 and GOST 31870 applications to determine of lithium

P A Volkov, S K Belus, V M Retivov, V V Zhavoronkov

Institute of chemical reagents and high purity chemicals, National research center "Kurchatov Institute", Moscow, 107076, bl. 3, Bogorodskiy Val, Moscow, Russia

e-mail: Volkov.Pavel.msu@yandex.ru

Abstract: The authors propose changes to existing standards GOST K 57165 and GOST 31870 in part of sample preparation for analysis on determining the lithium content in water by atomic emission spectrometry with inductively coupled plasma (AES-ICP)

Currently, high-tech control methods such as atomic emission spectrometry with inductively coupled plasma (AES-ICP) are increasingly used in elemental analysis. This method solves a wide range of analytical problems related, in particular, to environmental monitoring – analysis of drinking, surface, rain, mineral and wastewater. During the last ten years, the quality standards GOST R 57165 (ISO 11885:2007 "Water. Determination of elements by inductively coupled plasma atomic emission spectrometry") and GOST 31870 "Drinking water. Determination of elements content by atomic spectrometry methods" have been developed and realized in analytical laboratory practice. The standards have sufficient elaboration in the analysis of individual elements and their groups, but, unfortunately, do not take into account the behavior in the spectra of AES-ICP of some elements in the real matrix. One of the topicalities is the lithium content in water, since this element is often an indicator of industrial pollution.

The purpose of this work was to identify systematic patterns in determining the concentration of lithium in drinking water by AES-ICP in accordance with GOST R 57165 and GOST 31870.

Cold water of the Central water-supply of Moscow and model systems was used for measurements . In the study was realised methods using the method of fixed additives. The systematic effect of overestimating the readings of lithium content up to 60% with respect to the amount of the additive was revealed, that extended beyond the limit of the absolute error of the technique. It was shown that the response of lithium when using a wavelength of 670.78 nm is linear only with a constant composition of matrix elements, above all calcium, magnesium, sodium, and potassium, and grows nonlinearly with increasing their concentration. Using the recommended method for constructing a calibration dependence in accordance with GOST R 57165 and GOST 31870 can become the cause of an invalid data.

The detected effect makes it impossible to use these methods in interlaboratory comparisons in the case of certification of control samples according to the sample preparation procedure. A solution that allows analytical laboratories to reduce the systematic error and increase the accuracy of the analysis may be to emendation the currently GOST, permitting to use of the additive method in determining the contents of some elements.

Light Air Ions Generators and Air Ion Concentration Working Standards

N Nechaev, S Kolersky, P Zubkov

All-Russian Scientific Research Institute of Physicotechnical and Radio Engineering Measurements (VNIIFTRI), Mendeleevo, Moscow region, Russia, 141570

e-mail: mera@vniiftri.ru

Abstract: The article describes the requirements for air ion generators used in air ion concentration working standards and gives the results of the tests of air ion generators designed in accordance with these requirements.

The unit of light air ion concentration in Russia is transferred in accordance with the State Calibration Chain GOST R 8.646-2008 [1]. According to this standard, working standards should consist of standard air ion counters and radionuclide air ion generators (AIG). The calibration chain [1] does not contain any requirements for AIGs in working standards. Because of this, the question of the transfer accuracy is essential in designing working standards. This report is an attempt to formulate such requirements to guide further developers of air ion concentration working standards.

For accurate transfer of the unit the AIGs must meet the following requirements:

(a) The AIGs must output air ion concentration values in ranges set forth in the verification method GOST R 8.844-2013 [2]. It is also necessary to make sure that the output values don't drift during the calibration period (1 year) and that there is no intraday fluctuation.

(b) The output values must be stable during at least 3 minutes (approximate time sufficient for taking consecutive measurements with the standard air ion counter and an air ion counter under calibration). Failure to meet this requirement may result in an increase of false negative results due to a large random error component.

(c) AIGs must generate bipolar ionization. It minimizes the auxiliary error coming from the bulk charge in the air. The method of evaluation of this error is described in [3].

For (a) and (b), it seems reasonable to set the following requirement: the standard deviation of observations must not exceed 1/8 of the maximum permissible error, i.e. $S \leq 2.5$ %.

These requirements need to be considered when developing air ion concentration working standards. Low stability of AIGs may result in an increase of false negatives or total inability to perform, while high unipolarity of the air on the outlet will raise the auxiliary error attributed to the bulk charge.

References

[1] **GOST R 8.646-2008** State System for Assurance of Uniformity of Measurements. State Calibration Chain for Measuring Instruments of Bulk Charge in Ionized Air and Air Ion Number Concentration
[2] **GOST R 8.844-2013** State System for Assurance of Uniformity of Measurements. Aspiration Air Ion Counters. Verification Method
[3] **P Zubkov, S Kolersky, S Kolerskaya** 2014 Evaluation of the effect of space charge on the accuracy characteristics of air-ionometric equipment // *Scientific Session of NRNU MEPhI-2014. Abstracts of the reports: in 3 vol.* **2** 186a

CRMs of surface density, thickness and mass fraction of tin-bismuth alloy coating on copper developed by UNIIM

V V Kazantsev, A S Vasilyev

Ural state research institute for metrology (UNIIM), 620000, Ekaterinburg, Russia

e-mail: kazantsev@uniim.ru

Abstract. Provided results of CRMs certification. Tested CRMs of surface density, thickness and mass fraction of tin-bismuth alloy coating on copper using the state primary measurement standard of surface density and mass fraction of coatings GET 168-2015.

UNIIM certified CRMs of surface density, thickness and mass fraction of tin-bismuth alloy coating on copper in 2018. The CRMs are copper cylinders coated with a tin-bismuth alloy; the coating thickness is from 4 to 24 µm. UNIIM certified not only value of the surface density and thickness of coating of the CRMs, but also mass fraction of tin and bismuth in the coatings of the CRMs.

UNIIM certified the values of the CRMs using the state primary measurement standard of surface density and mass fraction of coatings GET 168-2015 [1]. X-ray fluorescence method that is realized on the GET 168-2015 is a relative method [2]. We used reference materials to transfer the unit of mass fraction. We used the secondary measurement standard of mass fraction and mass (molar) concentration of metals in solid and liquid substances and materials GVET 196-1-2012 to measure the mass fraction of elements in the coating of the reference materials.

Certified values are shown in tables 1 and 2.

Table 1. Surface density and thickness of the CRMs coating

CRM	Surface density of coating, g/m^2	Thickness of coating, µm	Relative error (P=0.95), %
UNIIM PPTM-OV/M-1	21.2	2.91	± 2.5
UNIIM PPTM-OV/M-2	38.4	5.26	
UNIIM PPTM-OV/M-3	63.9	8.75	
UNIIM PPTM-OV/M-4	128.5	17.6	

Table 2. Mass fraction of elements in the CRMs coating

CRM	Mass fraction of tin		Mass fraction of bismuth	
	Certified Value, %	Relative error (*P*=0.95), %	Certified Value, %	Relative error (*P*=0.95), %
UNIIM PPTM-OV/M-1	98.68	± 3	1.32	± 9
UNIIM PPTM-OV/M-2	98.68		1.32	
UNIIM PPTM-OV/M-3	98.69		1.31	
UNIIM PPTM-OV/M-4	98.78		1.22	

References:

[1] Kazantsev V, Medvedevskih S, Vasilyev A 2018 *Measuring technique* **9**
[2] Kazantsev V, Vasilyev A 2018 *Reference materials* **14** 3-4

About the development of reference materials of meat and meat product composition

Medvedevskikh M Yu [1], Sergeeva A S [1], Krasheninina M P[1], Vostrikova N L[2], Semenova A A [2], Kuznetsova O A[2]

[1]Ural Scientific-Research Institute of Metrology (UNIIM), Yekaterinburg, Russia
Researcher ID: G-6171-2017, e-mail: medvedevskikh_m@uniim.ru
[2]Gorbatov Research Center for Food Systems, 26, Talalikhina Str., 109316, Moscow, Russia

Abstract. The paper describes the development of certified reference materials (CRM) of meat and meat product composition

Control of meat product quality is of great importance for the preservation and maintenance of human health. The nutritional value of meat products depends on the content of proteins, fats and carbohydrates. The content of moisture (solids) is also a significant quality indicator, which determines the consumer properties and storage stability of meat products. Proteins and fats have important physiological functions in the human body, especially during the growth period. In this connection, the allowable levels of mass fractions of protein, fat and solids in meat products are specified in the Technical Regulation of the Customs Union [1]. These indicators should be measured by validated methods of measurement using measuring instruments of approved types [2]. To control accuracy of measurement results, graduation and calibration of measuring instruments, the certified reference materials (CRMs) of meat product composition are necessary.

In 2018-2019, the UNIIM together with the Gorbatov Research Center for Food Systems developed and carried out validated tests for the set type of CRMs of sublimated meat product composition.

The CRM material is sublimated cooked minced meat (pork, beef or poultry) packed (20-100 g) in double hermetically sealed polyethylene or metallized bags.

Traceability of the certified values of mass fractions of moisture and nitrogen (protein) in the CRMs under development is ensured by the measurements using the following standards:

- State primary standard of units of the mass fraction and mass (molar) concentration of water in solid and liquid substances and materials GET 173-2017;
- State secondary standard of units of the mass fraction and mass (molar) concentration of components in solid and liquid substances and materials based on the volumetric titration method of analysis GVET 176-1-2010, which in turn has metrological traceability to the units of the values reproduced by the State primary standard of units of the mass (molar) fraction and mass (molar) concentration of a component in liquid and solid substances and materials based on coulometry GET 173-2017.

Traceability of the certified values of mass fractions of fat is ensured by the strict adherence to the procedure of measurements according to the State primary reference procedure for measurement of the mass fraction of fat in food and food raw materials M.241.01/RA.RU.311866/2018 (number FR.PR 1.31.2018.00001 in the Federal Information Fund for Ensuring the Uniformity of Measurements).

The CRM analyses were carried out according to GOST ISO Guide 35 [3] and included several stages: determination of heterogeneity, long-term and short-term instability of the CRMs material, assessment of the certified values, verify of CRMs applicability for the control of measurement results obtained by standardized procedures.

Metrological characteristics of the CRMs are presented in table 1.

Table 1. Metrological characteristics of the CRMs of the sublimated meat product composition (set MP-1 CRM UNIIM) GSO 11274-2019/ GSO 11276-2019

Attested characteristics	Range of certified values, %	Absolute error of certified value (at $P = 0.95$), $\pm\Delta$, %
Moisture mass fraction	0.20 – 4.00	0.08
	4.00 –10.00	0.20
Nitrogen mass fraction [1]	3.20 – 15.20	0.24
Protein mass fraction [1,2]	20.0 – 95.0	1.5
Fat mass fraction [1]	5.0 – 15.0	0.7
	15.0 – 40.0	1.0

[1] The values are given on oven-dry basis.
[2] The nitrogen-to-protein conversion factor is 6.25.

The quality system of CRM producer (UNIIM) has been verified by the regional metrological organization COOMET (Euro-Asian Cooperation of National Metrological Institutions) according to ISO Guide 34 [4].

The CRMs shelf-life is 6 months; the production form is serial production with periodically repeated lots.

Therefore, the joint efforts of the institutes allowed developing the CRMs can be used in the field of legal metrology, as well as for calibration of measuring instruments for meat product composition and different kinds of express analyzers. This is especially topically in the laboratories of meat industry enterprises.

References.

[1] TR CU 034/2013 "On safety of meat and meat products": Technical Regulation of the Customs Union No. 68 of October 9 2012
[2] Federal Law “On technical regulation” No 184-FZ of December 27 2002
[3] GOST ISO Guide 35-2015 Reference materials. General and statistical principles for certification
[4] ISO Guide 34:2009 General requirements for the competence of reference material producers

Journal of Physics: Conference Series **1420** (2019) 012031 doi:10.1088/1742-6596/1420/1/012031

The choice of material for the creation of reference materials based on crystalline hydrates of alkali and alkaline earth metal salts

Anna Sergeeva, Anastasia Kasiliunas, Mariya Medvedevskikh, Sergei Medvedevskikh

Ural Scientific Research Institute for Metrology (UNIIM), Krasnoarmeiskaya St. 4, 620000 Ekaterinburg, Russian Federation

e-mail: lab241@uniim.ru

Abstract. The paper describes a study of chemically stable materials based on crystalline hydrates of alkali and alkaline earth metal salts for the subsequent development them as RMs

The State verification schedule for measuring of water content of firm and loose materials was developed and approved in accordance with the Order of the Federal Agency for Technical Regulation and Metrology No. 2832 on December 29, 2018. It establishes the procedure for transferring units from the State primary measurement standard of mass fraction and mass (molar) concentration of water in liquid and firm substances and materials (GET 173) to working standards and measurement instruments.

In the approved verification schedule, the transfer of units of water content from the GET 173 to the measuring units is carried out using reference materials (RMs) [1, 2]. By now there is a limited range of reference materials of approved types for the transfer of units of water content. But use of these RMs is not appropriate, in particular the organization and participation in key, pilot and additional comparisons of national primary measurement standards are required. The reason is matrix effects in most existing RMs. The scope of RMs is limited by strict compliance to regulated procedures, the accuracy of establishing certified values is insufficient, and such RMs have low shelf life.

In this regard, it is relevant to search for chemically stable materials based on crystalline hydrates of alkali and alkaline earth metal salts for the subsequent development of RMs.

The information on the physicochemical properties of crystalline hydrates of salts based on alkali and alkaline earth metals and knowledge of their heating behavior (heat resistance) is fundamental in the development of RM. The crystalline hydrates are the crystals containing water molecules. The formation of these crystals is due to the formation of stronger cationic bonds with water molecules in the crystal lattice in comparison with bonds between cations and anions in an anhydrous substance crystal. Such materials can contain a different number of water molecules - from less than one molecule (for example, calcium sulfate hemihydrate $CaSO_4 \cdot 0.5H_2O$) to twenty water molecules (for example, potassium metatungstate octadecahydrate $K_6H_2W_{12}O_{40} \cdot 18H_2O$), which provides a wide range of measurements of water content.

It is known that, one of the interfering factors in characterizing RM is water adsorbed from air according to the Rebinder classification [3]. At the same time, it should always be taken into account that the stop of the decrease of the mass of the sample during heating does not always reflect the complete removal of moisture from it. Therefore, it is necessary to take into account the effect of the residual water content in establishing of the certified value of RM.

To search for candidate materials for the creation of RM, work was begun to systematize data on the number of water molecules and the temperatures of its removal to an anhydrous state in crystalline hydrates available in various references [4-6]. In order to clarify the literature information, additional studies were carried out using a reference installation based on thermo-gravimetric analysis with mass-spectrometric detection (TGA-MS) as a part of the state primary standard GET 173.

The first RM developed by "UNIIM" on the basis of crystallohydrate was RM of the mass fraction of water in the dihydrate sodium molybdate (GSO 10911-2017). The range of permissible values of the mass fraction of water is from 14.70 % to 15.00 %, extended uncertainty at k=2 - 0.10 %. The main advantages of GSO 10911-2017 are the constant content of chemically bound (crystallization) water and the absence of matrix effects.

Another promising material for the creation of RM with improved stability characteristics is calcium oxalate monohydrate. The temperature stability of calcium oxalate monohydrate is due to the strength of water bonds with the skeleton of the salt forming the crystallohydrate. Thus, if crystallization water is retained by Van-Der-Waals forces, such substances are stable at temperatures below zero. In calcium oxalate monohydrate, crystallization water is retained by weak intermolecular bonds, so it is easily removed during heating. The low decomposition temperature (200° C) provides the convenience of using of this substance for the calibration of universal water content analyzers.

Thus, by selecting crystallohydrates with sufficient stability and homogeneity and having a different number of water molecules, a "line" of RMs in a wide range of water mass fraction values from 3.0 % to 50.0 % can be created. The results of studies obtained in the selection of materials-candidates for RMs can also be used to create a database of thermograms of crystallohydrates, in the long term issued in the form of a document of the state system of standard reference data (GSSD) [7].

References.

[1] M Yu Medvedevskikh et al. 2019 Creation of transfer measurement standards for implementation of State verification schedule for measuring of water content *Measurement Techniques* **6** 3-10, DOI: 10.32446/0368-1025it. 2019-6
[2] S V Medvedevskikh et al. 2015 General Approaches to the Estimation of Uncertainty in the Results of Reproducing Units of Water Content in Solids and Materials *Measurement Techniques* **58** (8) 926-933
[3] Rebinder P A 1966 Physico-chemical mechanics of dispersed structures (Moscow: Science) 399
[4] Lurie Y Y 1971 Handbook of analytical chemistry 4th ed.; processed and supplemented (Moscow: Chemistry) 456
[5] Knunyants I L 1988-1998 Chemical encyclopedia (in five volumes) (Moscow: Soviet encyclopedia) Great Russian encyclopedia **1-5**
[6] Ryabchikov D I 1966 Analytical chemistry of rare earth elements and yttrium (Moscow: Science) 381
[7] GOST R 8.614-2018. State system for ensuring the uniformity of measurements. State service of standard reference data. Basic principles (Moscow: Standartinform)

Development of reference materials with the composition of propyl and isopropyl alcohol solutions

I A Piterskikh[1], S V Vikhrova[1], O L Rutenberg[1], N G Kovaleva[2], T O Barinskaya[3]

[1]All-Russian Research Institute for Metrological Service, Moscow, tel.: +7(495)781-86-39; e-mail: piterskikh@vniims.ru
[2]«ECROSKHIM» LTD, Saint-Petersburg, tel. +7(812)448-28-37; e-mail: kovaleva@ekroschim.ru
[3]Public Healthcare Institution «Moscow scientific practical centre of addictology», Moscow, tel.: +7 (495) 660-20-56; (499)619-6049; e-mail: barto1605@yahoo.com

Abstract. In this work the development of RMs of mass concetration of the main components (propanol-1 and propanol-2) in biological objects (urine, blood) are described.

With the development of reference measurement procedures of mass concentrations of toxic substances (alcohol) in biological objects (urine, blood), it became necessary to create reference materials (RM) composed of propyl and isopropyl alcohol solutions for calibration of chromatographs, validation of measurement procedures and control of measurement errors of measurement results.

The characteristic of RM, used for these purposes, that is under the validation, is the mass concentration of the analyte, in this particular case, propyl and isopropyl alcohols. At the same time, the relative error of the certified RM values should not exceed 2% in order to ensure the required relative error of the measurement results of the mass concentration of these alcohols in water, urine and blood when measuring according to the developed reference procedure.

Metrological characteristics of solutions of pure substances are most often established by calculable and experimental method according to the manufacturing procedure. The traceability of the certified values in this case is ensured by the use of verified measuring instruments (MI) and the initial substance with metrological characteristics established in accordance with [1].

To ensure the rated values of metrological characteristics of RM with the composition of propanol and isopropanol solutions, a manufacturing procedure has been developed for each type of RM, which regulates the entire technological process of RM production: requirements for the initial substances, their preparation, storage conditions, RM manufacturing process, packaging, transportation and storage.

The peculiarity of manufacturing the solutions of volatile liquid substances in contrast to solids is the difficulty of taking a sample of a certain mass, which was eliminated by using specially made

Journal of Physics: Conference Series **1420** (2019) 012032 doi:10.1088/1742-6596/1420/1/012032

capillary pipettes. The optimal manufacturing procedure for the RM was chosen from among three options obtained by variation of the size and volume of the canopies.

The correctness of the CRM preparation procedure was confirmed by establishing the traceability of the certified RM values to comparison standards Xd 2.8400.029-12 / C3H8O-1 - 2018 and Xд 2.8400.029-13 / C3H8O-2 - 2018 (VNIIM named after D.I. Mendeleev) by the method of vapor-phase chromatography in accordance with MI 3174-2009 [2]. The certified characteristic of the comparison standards are the mass fractions of the main components, propanol-1 and propanol-2, respectively.

The certified values of RM established by the manufacturing procedure and the comparison standards coincided within the measurement error.

Materials of the development of RMs with the composition of propyl alcohol solutions have been forwarded to the State Service for Standard Specimens Relating to Composition and Properties of Substances and Materials (GSSO) (UNIIM) for the type approval of state CRMs.

References.

[1] RMG 55-2003 GSI (State system for assurance of measurement uniformity). Certified reference materials of pure organic matter composition. Methods of certification. Basic

[2] MI 3174-2009 GSI. Recommendation. GSI. Establishment of traceability of certified values of reference materials

Measurements of sulfur, phosphorus and nickel in biological samples like bovine liver by atomic emission spectroscopy

A A Stakheev, T P Stolboushkina

Russian Research Institute of Physicotechnical and Radio Engineering Measurements (VNIIFTRI), Moscow region, Mendeleevo, 141570, Russia

E-mail: stakheev@vniiftri.ru

Abstract. The work considers the method of atomic emission spectroscopy, the preparation of laboratory plasticware, the preparation of samples, the parameters of the spectrometer and processing measurement results are given. As confirmation of the correctness of the measurements, the results of the international comparison CCQM-K145 are presented.

Liver plays a major role in metabolism and acts as a source of energy for the body by storing glycogen. Working with other systems and organs, the liver is also responsible for several important function such as storing iron, detoxifying harmful substances, maintaining the hormonal balance and producing immune factors to fight infections. Cattle, like a human, seemed to be the most sensitive animal species with respect to some toxicities resulting from ingestion of feed material. With the growing interest in the biological effects and also for the meet measurement quality, it is important and necessary to develop accurate and comparable analytical methods for elements in bio-samples.

Inductively coupled plasma atomic emission spectroscopy is one of the most accurate methods of physicochemical analysis. The standard additions (solutions of the corresponding elements prepared from SRMs traceable to NIST, with concentrations 20, 40, 60, 80 and 100 % of the previously measured by external calibration) atomic emission spectroscopy method was used for the measurements. During the all performed measurements, to reduce the concentration of target elements in the "blank" sample and minimize the risks of contamination, the procedure for treatment of high density polyethylene plasticware, pipette tips and centrifuge tubes in an ultrasonic bath in four stages using diluted distilled nitric acid and deionized water was carried out, and treatment of perfluoroalkoxide plasticware in six stages with the additional use of concentrated distilled nitric acid was made as well. To increase the accuracy of the analysis, the following was carried out: all solutions were prepared gravimetrically; in all measurements the internal standard was used; to select the optimal concentrations of additions, a preliminary measurement of the target elements concentrations was carried out by the external calibration. The parameters of the spectrometer are shown in the table 1. Processing of the results was carried out with 5 pixels with dynamic background fit and x-y errors of the calibration curve.

The described method was tested during measuring of sulfur, phosphorus and nickel in the international key comparisons CCQM-K145.

The results of comparisons (Figures 1-3) within the reported uncertainties [1] are consistent with the reference values of the international comparison, which proves the applicability of the standard additions atomic emission spectroscopy method for measuring sulfur, phosphorus and nickel in bio-samples.

Table 1. The parameters of the spectrometer

Parameter	Value
Nebulizer	Concentric 1 ml per min
Spray chamber	Cyclone type
Cooling gas flow	12,0 L per min
Plasma gas flow	0,5 L per min
Nebulizer gas flow	0,6 L per min
Power RF-generator	1,2 kW
Integration time	10 s
The number of measurements per sample	5

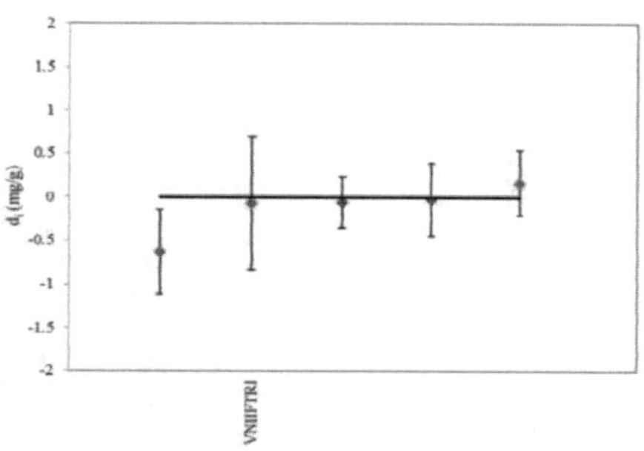

Figure 1. Equivalence statement of sulfur for CCQM-K145

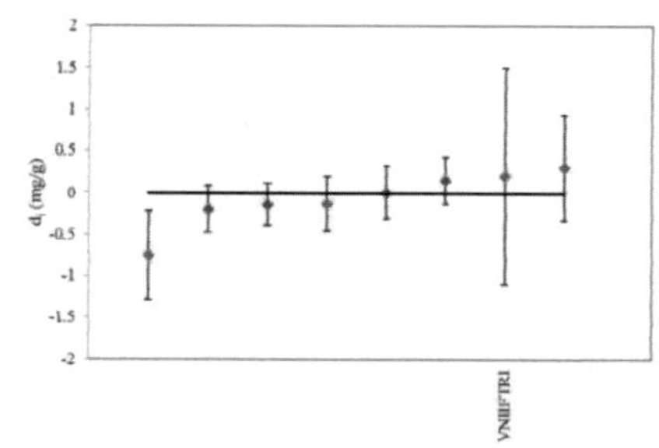

Figure 2. Equivalence statement of phosphorus for CCQM-K145

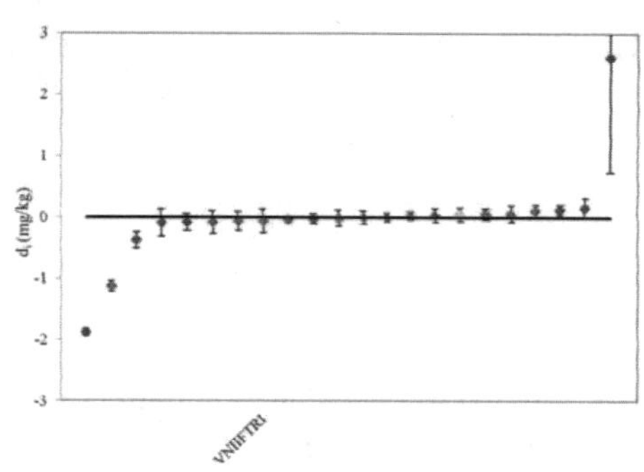

Figure 3. Equivalence statement of nickel for CCQM-K145

Based on the results of comparison, we are planning to submit an application for 3 calibration and measuring capabilities from the Russian Federation.

References

[1] CCQM GUIDANCE NOTE: Estimation of a consensus KCRV and associated degrees of equivalence *Version 10* 2013

Journal of Physics: Conference Series **1420** (2019) 012034 doi:10.1088/1742-6596/1420/1/012034

Development of the Certified Reference Materials in the field of physical and chemical analysis. Reference Materials of the composition of aqueous solutions of yttrium and scandium ions

I I Ermakova, A V Ivanov, I N Ziablikova, A N Shobina

All-Russian Scientific Research Institute of Optical and Physical Measurements (VNIIOFI), Moscow, Ozernaya 46, Russia

e-mail: vniiofi@vniiofi.ru

Abstract. In this work is described to study on rare-earth elements scandium and yttrium were, the development of certified reference materials of aqueous solutions of yttrium and scandium ions.

In modern industry, the application and production of rare-earth metals (REM) is actively developing, due to the need for new highly efficient materials. The mineral resource reserves of rare and rare-earth metals in Russia take a one of leading place in the world. The quaity of domestic deposits in the content of useful components, as well as a manufacturability, mining and processing conditions and availability not so good. At first, to prepare a rare-earth concentrate from crude ores, then it is sent to separation of individual elements.

In this work rare-earth elements scandium and yttrium were studied on.

Scandium (Sc) is mainly used to increase the strength of aluminum alloys, since has the same density with aluminum, but having a much higher melting point. Aluminum-scandium alloys are used in aviation and aerospace rocket production, as well as for the production of automobile engines. An important area of scandium oxide application is the computer ferrites production. Such ferrites less heat during magnetization reversal, which increases the speed of the computer's magnetic memory. Scandium hydride is used in nuclear engineering as a high-temperature neutron moderator.

Yttrium (Y) is widely used in ferrous and non-ferrous metallurgy; it increases the electrical conductivity of aluminum, the heat resistance of nickel-chromium alloys, the hardness and wear resistance of cast iron, and improves the mechanical and magnetic properties of steel. Yttrium has high tensile strength and melting point; therefore, it is able to replace titanium in any its application. Yttrium alloys are used in the aerospace industry, nuclear power plants, the automotive industry and in gas-phase rocket engines.

According to the Federal Law of January 10, 2002 No. 7-FZ "On Environmental Protection", the production of rare-earth metals is included in category I, which is mean that such enterprises have negative influence on the environment. Modern measuring instruments play an important role in this field [1]. So, the development of the reference materials for the modern measuring instruments to control the environmental pollution and rational use of the resources have been actual in Russia.

VNIIOFI has carried out the development of reference materials for the composition of aqueous solutions of yttrium and scandium ions in order to ensure traceability of measurements of these elements.

At first, foreign and domestic reference materials of the elements selected by VNIIOFI was analyzed.

Characteristics of the domestic reference materials are shown in Table 1. Characteristics of the foreign reference materials are shown in Table 2.

Table 1. Characteristics of the domestic reference materials.

Registration number	Name	The certified characteristic (mass fraction),%	Accuracy of the certified value (at P = 0.95),%
GSO 6649-93, validity period of the UTSO certificate until 2012.	Standard sample of the mass fraction of yttrium in a solid base (SRM-165)	1,00	0,02
GSO 6604-93, validity period of the UTSO certificate until 2012	Standard sample of the mass fraction of scandium in a solid base (SRM-89)	1,00	0,03

Table 2. Characteristics of the foreign reference materials.

Number and manufacturer	Name	Certified performance (mass concentration)	Extended uncertainty (for k = 2)
40523 «Sigma-Aldrich Production GmbH»	Yttrium Standard for AAS (Yttrium Oxide)	1000 mg / l	±6 mg / l
68418 «Sigma-Aldrich Production GmbH»	Scandium Standard for AAS (Scandium Oxide)	1001 mg / l	±6 mg / l
SRM 3167a, NIST	Standard solution of yttrium (yttrium oxide)	9,993 mg / g	±0,025 mg / g
SRM 3148a, NIST	Standard solution of scandium (yttrium oxide)	9,969 mg / g	±0,030 mg / g

Our reference materials of the aqueous solutions of yttrium and scandium ions were developed by dissolving chips of high-purity rare-earth metals in 3M nitric and 1M hydrochloric acids. In the manufacture of the reference materials was used Russian raw materials:

- yttrium metal grade ItM-1 according to TU 48-4-208-72 with a mass fraction of yttrium of at least 99.8%
- scandium metal distillate of the SKM-1 brand according to TU 48-4-483-87 with a mass fraction of scandium of at least 99.992%.

The mass fraction of the main component in them was calculated as the difference between 100% and the total content of impurities. The determining the purity of the carrier of the certified element was carried out on a Quantum-Z atomic absorption spectrometer. ETA-T (Kortek LLC, Moscow), which is part of the GET 196-2015. The values of the mass concentration of yttrium and scandium ions were calculated according to the preparation procedure and then confirmed by measuring on an optical emission spectrometer with inductively coupled plasma Ultima 2 (Horiba Jobin Yvon), which is a part of the Steat Prymary Standard GET 196-2015. The procedure measurements

FR.R1.31.2019.00003 "Reference method for measuring the mass fraction and mass concentration of rare-earth elements in alloys based on aluminum and nickel" was used.

The studu of the instability of the reference materials was carried out in accordance with RMG 93-2015 “State system for ensuring the uniformity of measurements. Assessment of the metrological characteristics of the raference materials ”by the isochronous method [2].

As a result of the studies values of mass concentration of the rare-earth metal ions and expanded measurements uncertainty were obtained (Table 3).

Table 3. Values of mass concentration of the rare-earth metal ions and measurements uncertainty

Name	Mass concentration g / dm^3	Extended uncertainty (at k = 2),%
Reference material of the yttrium ions solution	from 0.95 to 1.05	1,0
Reference material of the scandium ions solution	from 0.95 to 1.05	1,0

The VNIIOFI reference materials of the yttrium and scandium ions solutions will be produced in ampoules with valums of 5 and 10 cm3 and can be used for calibration and verification of measuring instruments (used methods: atomic absorption, atomic emission, photometric, X-ray fluorescence and others), testing of measiring instruments to type approval, accuracy control of measurement results, certification procedure.

References.

[1] Federal Law dated 10.01.2002 No. 7-FZ “On Environmental Protection” (as amended on 12/27/2018)

[2] RMG 93-2015 “State system for ensuring the uniformity of measurements. Evaluation of the metrological characteristics of standard samples” (Moscow: Standartinform) 2016 32

Journal of Physics: Conference Series **1420** (2019) 012035 doi:10.1088/1742-6596/1420/1/012035

Reference materials for inductively coupled plasma methods (ICP standards) produced by VNIIFTRI

T P Stolboushkina, A A Stakheev

All-Russian Scientific Research Institute of Physicotechnical and Radio Engineering Measurements (VNIIFTRI), Mendeleevo, Moscow region, Russia, 141570

E-mail: stolboushkina@vniiftri.ru

Abstract. For the metrological assurance of the high-precision inductively coupled plasma methods, the Russian Metrological Institute of Technical Physics and Radio Engineering (VNIIFTRI) developed four types of certified reference materials (CRMs) of the mass fraction of metals in solutions (ICP CRM). The permissible certified values of the element mass fraction (mass concentration) in the developed ICP CRM range from 800 mg/kg to 1200 mg/kg (from 800 mg/dm^3 to 1200 mg/dm^3). The developed ICP CRMs can be used for ensuring the metrological traceability of measurements in inorganic analyses to the State Primary Standard for units of the mass fraction and mass (molar) concentration of inorganic components in aqueous solutions based on gravimetric and spectral methods GET 217-2018.

1. Introduction

Currently, one of the most important tasks of science and industry is the improvement of measurements. Therefore, the role of chemical metrology is constantly growing, especially in the field of analytical chemistry. Due to technical progress, quality control of different materials is an increasingly important element of the creating safe, reliable products of high quality.

Inductively coupled plasma spectrometers are widely used for elemental analysis of more than 70 elements. These methods are the most express, highly sensitive methods for qualitative and quantitative analysis. Like many other methods of chemical analysis ICP methods need calibration. The ICP calibration procedure requires the employment of standards whose values and their associated uncertainties are known and established following a metrological procedure. The most rational way to achieve comparable and reliable measurement results by ICP-MS and ICP-OES consists in establishing the metrological traceability of measurements using CRMs. The accuracy of measurements a strongly depends on the quality of CRM and, therefore, their characterization and traceability is very important.

This paper presents the results of the development of four types of certified reference materials for inductively coupled plasma methods (ICP CRM) are developed in VNIIFTRI, their main features and differences. ICP CRMs produced by VNIIFTRI are a crucial link in the traceability chain of measurements to the State primary standard of mass fraction units and mass (molar) concentration of inorganic components in aqueous solutions based on gravimetric and spectral methods GET 217-2018.

2. Materials and methods

During the preliminary experiments the starting material was selected for each type of standard. In order to assess the purity of the starting material, the "100% minus the amount of impurities" method was chosen. The mass fraction of the main component (W) was calculated by equation (1):

$$W = 100\% - \sum_i w_i - \sum_j \frac{LOD_j}{2} \tag{1}$$

where

w is the mass fraction of detected impurities, %;

LOD is the limit of detection of undetected impurities, %.

HNO_3 were produced by double distillation of reagent grade acids through a quartz subboiling distillation unit. Deionized water was used for dilution of solutions. The impurity composition of the solvents (nitric acid and deionized water) was assessed on the GET 217-2018.

The main objective of the certification of a reference material is to determine its characteristics including their uncertainty.

The mass fraction of the certified component (A_m) in the reference material was estimated by equation (2):

$$A_m = \frac{\omega_1 \cdot m_1 + \omega_2 \cdot m_2}{m_1 + m_2} \tag{2}$$

where

ω_1, ω_2 are the mass fractions of the certified component in the starting material and solvent, respectively;

m_1, m_2 are the weights of the starting material and solvent, respectively.

The mass concentration of the certified component (A_v) in the reference material was estimated by equation (3):

$$A_v = A_m \cdot \rho \tag{3}$$

where

A_m is the mass fractions of the certified component in the reference material,

ρ is the density of the reference material.

The metrological characteristics of ICP standards was calculated in accordance with RMG 93-2015 Metrological characteristics evaluations of reference materials and GOST ISO Guide 35-2015 Reference materials. General and statistical principles for certification (In Russ.). The uncertainty components for the gravimetric preparation of CRM solutions are presented as a cause-and-effect diagram in figure 1.

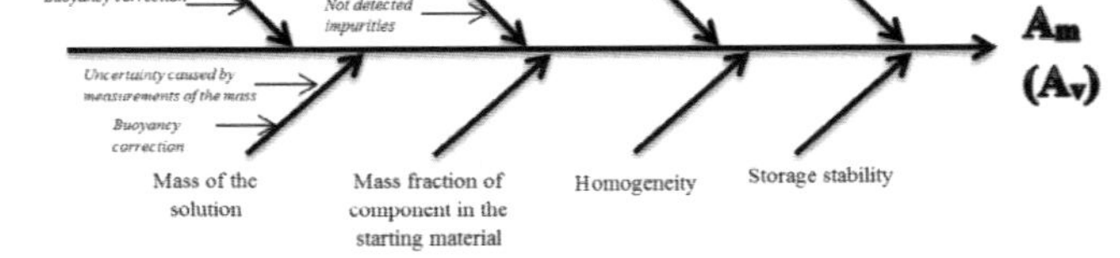

Figure 1. A cause-and-effect diagram uncertainty for uncertainty evaluation

3. Results and discussion

The reliability of the certified value fraction was established by two independent methods, including the analytical experimental method of gravimetric preparation and the method of measuring the analyte mass fraction using the capacity of the GET 217-2018. The measurement results obtained using both above mentioned methods are in good agreement within the limits of their uncertainties. The certified value of the developed CRM is taken to be that obtained by the analytical experimental method of gravimetric preparation. Metrological characteristics of the ICP CRMs are given in the table 1.

Table 1. The metrological characteristics of the ICP CRMs produced by VNIIFTRI

Product name	Element	Certified characteristics	Interval of certified values	Expanded uncertainty, U_o ($k = 2$; $P = 0{,}95$), %
Lithium standard for ICP (ICP CRM Li)	Li	Mass fraction, mg/kg Mass concentration, mg/dm^3	800 - 1200	0.5
Cobalt standard for ICP (ICP CRM Co)	Co			
Cadmium standard for ICP (ICP CRM Cd)	Cd			
Lead standard for ICP (ICP CRM Pb)	Pb			

As shown in table 1, the expanded uncertainty of the certified value is 0.5 %. Trace metallic impurities determined by ICP-OES and provided in the passport. Among other things, one of the main advantages of ICP CRM is the certified value of the mass fraction of the component in addition to the mass concentration. The characterization of the mass fraction of the component in the solution allows to use a more accurate method of preparation of calibration solutions. The introduction of the gravimetric preparation of solutions improves the quality of results from analytical measurements by reducing the uncertainty associated with the preparation of solutions. When using Class A volumetric flasks, the tolerances include up to 0.20 % limit of error. If transfer pipets are used, they can add up to an additional 0.60 % limit of error. This does not include the human error, or the influence of temperature changes. While the uncertainty of analytical balances is not more than 0.1 %, and most modern electronic analytical balances perform even better than this. The introduction of the gravimetric preparation of solutions presents a significant opportunity to improve the quality of results from analytical procedures by reducing the uncertainty associated with the preparation. In the case of an existing method, where the analyst should work with the mass concentration use, he also can use ICP CRM because of it have a second certified characteristic like a mass concentration of component.

ICP methods do not require a large volume of samples. Often 1 – 5 ml of solution is enough for analysis. The use of volumetric flask results in excessive use of expensive solvents and substances. Gravimetric sample preparation significantly reduces the amount of solvent and substance consumed and significantly reduces the amount of waste. The gravimetric sample preparation allow to use disposable laboratory plasticware, which will prevent the possibility of cross-contamination of samples from volumetric flasks.

4. Conclusions

Four types of CRM for ICP (figure 2) was successfully developed in VNIIFTRI. The homogeneity, stability, and uncertainty were sufficiently studied. ICP CRMs can be used for ensuring the metrological traceability of measurements in inorganic analyses by ICP methods. The certified value expressed in the mg/kg units allows a more accurate, gravimetric method of solution preparation to be used, whereas the certified value expressed in the mg/dm^3 allows, if necessary, a more familiar and fast volumetric method to be used instead.

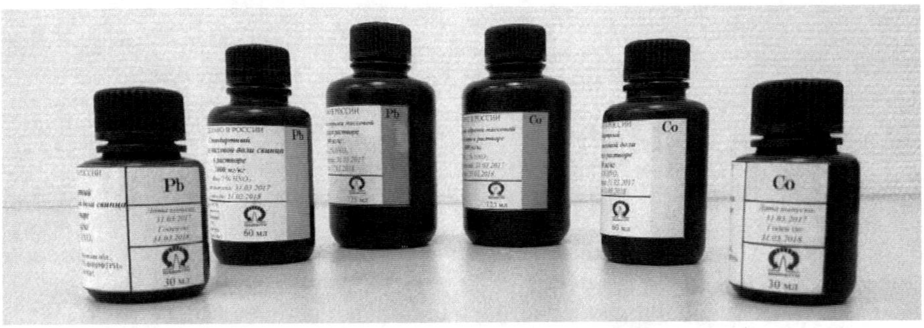

Figure 2. ICP CRMs developed in VNIIFTRI (ICP CRM Pb and ICP CRM Co)

REFERENCES.

[1] General Chapter <31> Volumetric Apparatus, USP 35-NF 30.
[2] RMG 93-2015 Metrological characteristics evaluations of reference materials
[3] GOST ISO Guide 35-2015 Reference materials. General and statistical principles for certification (In Russ.)

The Role of Distributions and Control Charts in Metrology

Yu Adler, T Polkhovskja, V Filichkina, V Shper

NIST "MISIS", Moscow, Leninskji prospect, 4

e-mail: vlad.shper@gmail.com

Abstract. In this paper it is suggested to change radically the existing metrology paradigm for measurement data analysis, namely to reject from searching the appropriate Distribution Function (DF) and use nonparametric methods for uncertainty estimation in all circumstances. This let to avoid using limiting and unprovable assumptions that lead to invalid conclusions. In the paper it is offered to use Shewhart Control Charts (ShCC) as an indispensable method to monitor the current state of measurement system. And this method does not require any assumptions about specific type of DF.

Papers on metrology give much attention to the role of Distribution Functions (DF) for measurement treatment and interpretation. However most often it is impossible to estimate which specific DF relates to a sample. Besides this requires big resources and much time. That's why practitioners use the idea of normality so permanently. More than two centuries it seemed quite acceptable to them. As a result many are used to consider it as something natural and obvious. Great reduction of cost for calculations and rapid growth of computer power and data volumes let people to create empiric estimates of sample distributions. Consequently, it turned out that something like normal DF is a very rare event. So we suggest here to change radically the existing metrology paradigm for measurement data analysis, namely to reject from searching the appropriate DF and use nonparametric methods for uncertainty estimation in all circumstances. While DFs are being discussed by metrologists often and with enthusiasm they talk about Shewhart Control Charts (ShCC) much more rare and – we think – not to the point. Meanwhile this is an indispensable method to monitor the current state of measurement system. And this method does not require any assumptions about specific type of DF. Besides we suggest using ShCC in order to check data homogeneity and to estimate widened uncertainty. Such procedures turn out to be much simpler than those described in many metrology standards.

Why normal distribution [1]? There are many reasons for metrologists to adhere to normal distribution. First of all, it is an old tradition sanctified by the name of outstanding Gauss. He was simply looking for an approximation to describe parallel observations of trajectories for astronomy measurements. But his grateful customers accepted his result as the great law of Nature, and they persuaded themselves in this. So all this started in 1806. This tradition was growing, widening, became more sound and successfully reached our days. And when you have the Law you do not need anybody, all you need are formulas to calculate and tables to simplify results and to accelerate the procedure. An useful additional result is that learning simple formulas and algorithm does not require notable time and efforts – this means that one can improve his mastering in his own profession where there are always many problems. But these are not all advantages.

It is not mandatory to know the theory which formulas are based on but anyone wants to be sure that these formulas give him the best result. And the theory claims that if you know that the DF is normal then these formulas give us the best optimal result. And if one uses some detour way – there are always many detour ways because statisticians need to earn for their living – the result will always be worse. This means that under the same efforts one will obtain wider confidence limits, the result will have greater uncertainty, its reliability will be lower, in short, everything will be much worse. Simultaneously efficiency of using resources will be suffering as well as waste of time will be significant. Some estimates show that average worsening will be about three times.

Obviously, the choice is clear. But there is a "little" difficulty. The thing is that all above-mentioned is absolutely true only under assumption that normality is unquestionable. Alas, the doubts do exist. And not little ones. In real life we are not interested in asymptotics and limiting theorems because we can't live until the time to use them will come. And the idea of normality at best cannot be rejected for real samples especially for little ones as well as many other ideas. That's why normality becomes the symbol of belief. It is dangerous to build important conclusions on such base.

Sometimes the practitioners are trying to use a priori information about some may be not normal distributions. Then one can estimate the uncertainty by the maximum likelihood method. It is clear that if we are sure that our data relate to, for example, Poisson DF, or Weibull, or lognormal distribution, then it will be wrong not to use this knowledge because it always will improve results and make them cheaper though not always this will make their extraction more simple. Unfortunately these cases are rare and exotic and do not change general picture.

If we can't rely on reliable information about DF maybe it is worth start to think about the estimates which do not afraid to break any assumptions. Such methods do exist and are called "robust" [2]. Their using is quite possible and is based on intensive use of computers and is difficult for interpretation. One can remember the robust estimate of mean, which was created by Tukey and was named "gopher estimate" by him. It pretends to be as sustainable to breaking assumptions as "gopher wood" – the material for Noath's Ark – had been during the Flood.

We think that the best choice for today is the approach, which admits that DF just exists for the random variable under consideration and which is limited with only this weak hypothesis. Such approach is usually named as nonparametric [3]. It is minimally pretentious. Simultaneously it let us to do everything we need and all calculations are much easier and interpretation is simpler.

What is the price for these advantages?

The price is that we cannot indicate the exact probabilities of measurement errors. But if we look at the exact values of uncertainties attentively we will understand that they emerge in classical metrology only under the hypothesis of normality, but we do not belief in it. That is the values of 95% or 99% are fictions, which mislead us. What for the exact values are necessary if we never can reliably estimate them? And rough estimate is always possible.

Why we need to use Shewhart Control Charts (ShCC) [4]? The application of ShCC encounters constantly on different misunderstandings. One of the reasons is that during monitoring the measurement process inside a process of manufacturing of some product we need to follow two interconnected processes. One is the production process itself and ShCC is necessary to find out the moment of the loss of stability in order to return the process into stable state. Another is the measurement process, which can also become unstable and needs to be returned into stable state. Measurement professionals usually used to consider that production process can go out from stable state. But they often think that measurement process cannot behave in such way. They consider it to be stable. But our experience shows that measurement processes can be unstable as often as production processes. Surely monitoring of measurement process should be organized separately from monitoring production process. To this end it is necessary to include into production process in randomized order the reference measures which then will be used for construction additional ShCC. The choice of the structure for this process, the frequency of including standard samples and other organization problems – all these are the ordinary tasks of design of experiment (DOE). The analysis of the corresponding chart will show if the loss of stability took place or not. And if this has happened we need to do everything we are doing for technological process. Namely, we need to create a team of professionals to search and remove the root causes of unstability, and this team should include those who measure as well as experts in metrology, statistics, equipment, economics, management. They need to learn to understand each other, i.e. to create common terminology, and become a team, that is to learn to work for their common goal while do not forgetting their professional purposes. Of course, this team must use the same tools as the team of technological process.

So, our suggestions may be presented as the following simple algorithm:

1. A rejection of searching the DF
2. The use of median (or any other nonparametric estimate of data centre) instead of mean
3. The use of sample range or interquartile range instead of sample variance (or sample error or sample standard deviation)
4. Visualization of measurement data by using the Tukey's "box-and whisker" and corresponding analysis of data for outliers [5]
5. Construction of histograms
6. Monitoring of measurement process by using ShCC (which do not require the normality)
7. The analysis of homogeneity of measurement data by using the rules for interpretation of ShCC
8. The estimate of measurement process stability
9. The creation of team for searching the root causes of unstability, and if it is found
10. Continuous improvement (kaizen) [6] of measurement process by using the Shewhart-Deming cycle (PDSA)
11. The estimate of widened uncertainty as the distance between the upper and lower limits of ShCC for the process in stable state
12. The search for new possibilities to improve measurement process by using better equipment, new technologies, and so on.

In our opinion there is nothing above that contradicts the requirements of ISO 17025 standard and ISO 9000 standards.

As a whole the thing is that it is high time to change the paradigm of that section of science which is called metrology. It is well and long ago known that paradigm change is a difficult and tiresome process taking usually a big period of time. We understand that our suggestions will be - most probably - rejected or admitted partly. So at this moment we suggest to organize a broad discussion about all above-mentioned problems and discuss not the abolition of traditional approach but only the possibility of application the suggested approach in parallel with traditional procedures. Surely, we are ready to participate actively in such discussion and work. The examples of analysis of real data and comparison with traditional approach are presented in our papers [7].

References.

[1] Adler Yu 2013 The Statistics: Past, Present, and Future (A New Paradigm in Statistics) *Proc. 13th Annual ENBIS Conference* (Ankara, Turkey)

[2] Huber P J 1981 *Robust Statistics* (NY, John Wiley & Sons)

[3]. Adler Yu, Shper V 2019 Practical Guide for Statistical Process Control (Moscow: Alpina Publisher)

[4] Mosteller F, Tukey J 1977 Data Analysis and Regression (USA:Addison-Wesley Publishing Company)

[5] Kume H 1985 Statistical Methods for Quality Improvement AOTS

[6] Adler Yu, Shper V 2019 Assessment in an Era of Changes: Is There Any Use to Change? Part 1: Why Does Uncertainty Need Normality? *Product Quality Control* **9** 30-36

[7] Adler Yu, Shper V 2019 Assessment in an Era of Changes: Is There Any Use to Change? Part 2: Analysis of Results of Interlaboratory Comparisons *Product Quality Control* **10** 27-35

Journal of Physics: Conference Series **1420** (2019) 012037 doi:10.1088/1742-6596/1420/1/012037

Capabilities of VNIIIFTRI in the field of measurements of meteorological parameters

N M Justus, V I Dobrovolskiy, K I Dobrovolskiy

All-Russian Scientific Research Institute of Physicotechnical and Radio Engineering Measurements (VNIIFTRI), Mendeleevo, Moscow region, Russia, 141570

e-mail: mera@vniiftri.ru

Abstract: The creation of new working standards at the All-Russian Scientific Research Institute of Physicotechnical and Radio Engineering Measurements (VNIIFTRI) for metrological support of meteorological measuring instruments is mentioned in the work.

Improving the accuracy of the State primary standards, as well as the working standards, requires improving the accuracy of measurements and maintaining the temperature and humidity of environment where these standards are located. In addition, modern high-precision production (microelectronics, mechanical engineering, modern chemical and biochemical technologies, etc.), calibration, testing and analytical laboratories require continuous monitoring of environmental parameters. Special automated and automatic weather stations are needed for the prevention of cataclysms and weather forecasting. All these factors furthered to increase the types and number of measurement instruments (MI) of meteorological parameters such as: air temperature, relative humidity, air velocity and direction, atmospheric pressure, rainfall intensity and amount, for the metrological support of which there increased necessity of the creation of the new working standards.

As example, technical parameters of some thermohygrometers used in Russia are shown in Table 1.

Table 1. Technical parameters of some thermohygrometers used in Russia

SI	Range of measurement	Absolute measurement error
Thermohygrometer IVA-6	from 0 to 98 %	±2 %
	from -50 to 180 °C	±0,3 °C
Digital thermometers small size TCM 1500	from -50 to 200 °C	±0,5 °C
Humidity and temperature meter IVTM-7	from 0 to 99 %	±2 %
	from -45 to 150 °C	±0,2 °C
	from 840 to 1060 hPa	±3 hPa

As example, technical parameters of some thermoanemometers used in Russia are shown in Table 2.

Table 2. Technical parameters of some thermoanemometers used in Russia.

SI	Range of measurement	Absolute measurement error
Thermoanemometer TTM-2	from 0,1 to 30 m/s	±(0,05 + 0,05·V), where V is the air flow velocity m/s
The anemometers TKA-SDV	from 0,1 to 20 m/s	±(0,1 + 0,05·V), where V is the air flow velocity m/s
Thermoanemometers TAM-M1	from 0,1 to 5 m/s	±(0,02 + 0,1·V), where V is the air flow velocity m/s

As example, technical parameters of some weather stations used in Russia are shown in Table 3.

Table 3. Technical parameters of some weather stations used in Russia

SI	Range of measurement	Absolute measurement error
Meteorological station HM-30	from -20 to 60 °C	±3 °C
	from 0 to 100 %	±2,5 (0 – 10) %; ±1,5 (10 – 90) %; ±2,5 (90 – 100) %
	from 225 to 1125 hPa	±1 hPa
Weather stations automatic WXT530	from -52 to 60 °C	±0,5 (from -52 to 20 °C); ±0,3 (from 20 to 40 °C); ±0,4 (from 40 to 60 °C).
	from 0 to 100 %	±3 (1 – 90) %; ±5 (90 – 100) %
	from 600 to 1100 hPa	±0,5 hPa
	from 0,2 to 60 m/s	±0,5 (from 0,2 to 10,0) m/s; relative error±5 % (from 10 to 60 m/s)
	Airflow direction 0 to 360 degree	±3 degree
	Minimum measured rainfall 0,2 mm	±(0,02 + 0,05·M), where M is the measured amount of precipitation, mm

VNIIFTRI is the holder of the State primary standard of the unit of relative humidity of gases located in the East Siberian branch of VNIIFTRI in Irkutsk. As a result, of growing the needs, the actual task

Journal of Physics: Conference Series **1420** (2019) 012037 doi:10.1088/1742-6596/1420/1/012037

was to create a working standard for unities of relative humidity of gases at the head institute of VNIIFTRI (Moscow Region, Solnechnogorsk, Mendeleevo).

Figure 1 and 2 show the portable wind tunnel PAT-60 (air flow rate from 0.1-60 m/s) and the climate test chamber Heat-Moisture-Cold M-70/150-1000KTVH (air temperature playback range -70 to 150 ºC) which have been set up at VNIIFTRI.

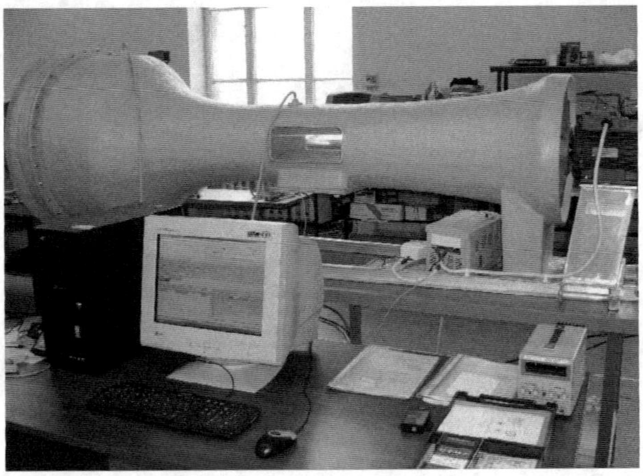

Figure 1. General view of the portable wind tunnel PAT-60 (air flow rate from 0.1-60 m / s), VNIIFTRI

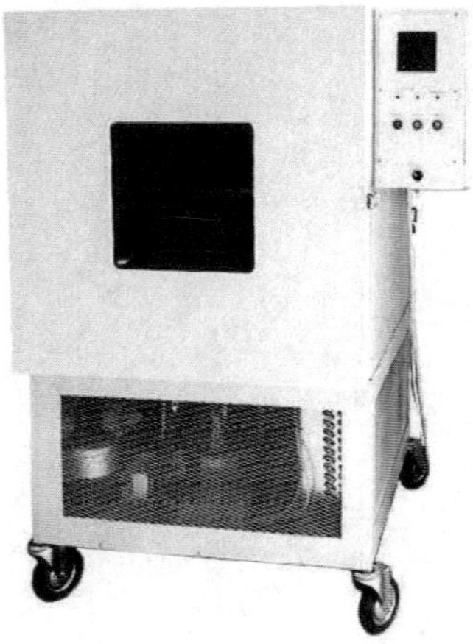

Figure 2. General view of the climate test chamber Heat-Moisture-Cold M-70/150-1000KTVH (air temperature playback range -70 to 150 ºC), VNIIFTRI

About 1000 pcs the measuring instruments (MI) were tested to approve the types and verified by the new working standard, due to solved this task. At the same time, working standards were created at the head institute of VNIIFTRI: 3.1.ZZT.0284.2018 – air temperatures in the range from minus 70 °C to 150 °C (instead of the previous working standard in the range from 0 °C to 60 °C); 3.1.ZZT.0319.2019 – Oatmospheric pressure from 500 hPa to 1100 hPa; 3.1.ZZT.0328.2019 – fluid volume from 0 cm^3 to 250 cm^3; 3.1.ZZT.0219.2015 – air-flow rate from 0.1 m / s to 60 m / s; 3.1.ZZT.0282.2018 – flat angle (air-flow direction) in the range from 0º to 360º.

Thus, the new working standards was developed at the VNIIFTRI that allow the metrological support of about all kind of meteorological MI. On the whole, the number of verified and tested MI in 2018 at VNIIFTRI, increased by 163 % compared to 2014, due to the developed new working standards.

AUTHOR INDEX